Lecture Notes in Mathematics

Edited by A. Dold and B. Eckmann

1125

Dynamical Systems and Bifurcations

Proceedings of a Workshop held in Groningen
The Netherlands, April 16–20, 1984

Edited by B. L. J. Braaksma, H. W. Broer and F. Takens

Springer-Verlag
Berlin Heidelberg New York Tokyo

Editors

Boele L. J. Braaksma
Hendrik W. Broer
Floris Takens
Rijksuniversiteit Groningen, Subfaculteit Wiskunde en Informatica
Landleven 5, 9747 AD Groningen, The Netherlands

Mathematics Subject Classification (1980): 58-xx, 34 C xx

ISBN 3-540-15233-4 Springer-Verlag Berlin Heidelberg New York Tokyo
ISBN 0-387-15233-4 Springer-Verlag New York Heidelberg Berlin Tokyo

Printing and binding: Beltz Offsetdruck, Hemsbach/Bergstr.
2146/3140-543210

PREFACE

These are the proceedings of the International Workshop on Dynamical Systems and Bifurcations, organized by the Department of Mathematics of Groningen University from April 16 to April 20, 1984. Apart from Groningen University financial support was given by various instances: Koninklijke Shell B.V., de Koninklijke Nederlandse Akademie van Wetenschappen and the Netherlands Organization for the Advancement of Pure Research (ZWO).

The informal atmosphere among the participants of the Workshop made it possible that there was an intensive exchange of mathematical ideas.

We are grateful to all who contributed to the success of our Workshop, in particular to the lecturers and to the authors of these proceedings.

B.L.J.B., H.W.B., F.T.
Groningen, march 1985.

TABLE OF CONTENTS

AN ELEMENTARY PROOF OF THE CONLEY - ZEHNDER THEOREM IN SYMPLECTIC GEOMETRY.

Marc CHAPERON

Centre de Mathématiques
Ecole Polytechnique
91128 Palaiseau Cedex (France)
"U. A. du CNRS n° 169"

In [9], C.C. Conley and E. Zehnder proved a celebrated conjecture of V.I. Arnold [2] on the number of fixed points of a symplectic diffeomorphism of the 2n-torus $\mathbb{T}^{2n}$. It was then shown in [5] that their method could be used to establish a more general and geometric result on Lagrangian intersections in the cotangent bundle $T^*\mathbb{T}^n$, also conjectured by Arnold [1].

This note is an improved version of [7] ; we replace the natural but rather subtle analytic argument of [9]-[5] by an elementary but rather tricky geometric method, inspired by M. Morse's theory of geodesics -see [3] or [10].

It is a pleasure to thank the organizers of this meeting for the friendly and stimulating atmosphere they were able to create.

1. STATEMENT AND FORMULATION OF THE MAIN RESULT.

All mappings are assumed smooth (enough). Let M denote the n-torus $\mathbb{R}^n/\mathbb{Z}^n$. Recall that the *Liouville form* on the cotangent bundle T^*M is the 1-form λ such that, for every 1-form α on M (considered as a mapping $M \to T^*M$), the pullback $\alpha^*\lambda$ is α itself. Its exterior differential $d\lambda$ is a nowhere degenerate (exact) 2-form on T^*M, called the *standard symplectic structure of* T^*M.

An immersion $j : M \to T^*M$ is called *Lagrangian* (resp. *exact Lagrangian*) iff $j^*\lambda$ is closed (resp. exact). If j is an embedding, this is a property of j(M), which is then called a *Lagrangian* (resp. *exact Lagrangian*) *manifold*.

MAIN THEOREM [5] .- *Let* $(j_t)_{0\leq t\leq 1}$ *be a family of Lagrangian embeddings* $M \to T^*M$, *depending smoothly on* t, *and with the following two properties* :

(i) j_1 *is exact* ;

(ii) $j_o(M)$ *is the zero section* Σ *of* T^*M.

Then $j_1(M) \cap \Sigma$ *contains at least* n+1 *points, and at least* 2^n *if* j_1 *is transversal to* Σ.

This result is trivial is the following two cases :

- If $n=1$, then the exactness condition just means that the (algebraic) area between the two circles Σ and $j_1(M)$ on the cylinder T^*M is 0.
- If the restriction of the canonical projection $T^*M \to M$ is a diffeomorphism $j_1(M) \to M$, then there exists a 1-form α on M such that $j_1(M) = \alpha(M)$, and, by the very definition of λ, the exactness condition means that α is exact. Now, by classical Morse and Lyusternik-Schnirelmann theory ([3], [10]), a function on M has at least n+1 critical points, and at least 2^n if none of them is degenerate.

Recall that an *isotopy* of a manifold X is a family $(\varphi_t)_{0\leq t\leq 1}$ of diffeomorphisms $X \to X$, depending smoothly on t, such that $\varphi_o = id_X$. If there exists a compact set $K \subset X$ such that $\varphi_t(x) = x$ for $x \in X \setminus K$ and $0 \leq t \leq 1$, then (φ_t) has *compact support*.

A symplectic version of J. Cerf's classical lemma on the extension of isotopies ([5] or [4]) yields

LEMMA 0 .- *There exists a compactly supported isotopy* $(\psi_t)_{0\leq t\leq 1}$ *of* T^*M *with the following two properties* :

(*i*) *for* $0 \leq t \leq 1$, *the* 1-*form* $\psi_t^*\lambda - \lambda$ *is exact* ;

(*ii*) $\psi_1(\Sigma) = j_1(M)$. □

Since (ψ_t) has compact support and satisfies (i), the following result is clear :

LEMMA 1 .- *There exist an integer* $N \geq 0$, *a sequence* $0 = t_o < t_1 < \ldots < t_{N+1} = 1$ *of real numbers, and functions* $f_o, f_1, \ldots, f_N : T^*M \to \mathbb{R}$ *such that the following hold true* :

(*i*) For $0 \leq i \leq N$, *if* $\varphi_i := \psi_{t_{i+1}} \circ \psi_{t_i}^{-1}$, *then* $\varphi_i^*\lambda - \lambda = df_i$; *therefore*, df_i *has compact support.*

(*ii*) *If* $\pi : T^*M \to M$ *denotes the canonical projection, then* $\pi \circ \varphi_{o|\Sigma} = \pi \circ \psi_{t_1|\Sigma}$ *is a diffeomorphism* $\Sigma \to M$.

(*iii*) For $1 \leq i \leq N$ *and* $y \in (\mathbb{R}^n)^*$, *the restriction of* $\pi \circ \varphi_i$ *to the horizontal space* $M \times \{y\} \subset M \times (\mathbb{R}^n)^* = T^*M$ *is a diffeomorphism* $M \times \{y\} \to M$. □

Thus by lemma 0 (i) and lemma 1, our problem is the following : *show that* $\Sigma \cap \psi_1(\Sigma) = \Sigma \cap \varphi_N \circ \ldots \circ \varphi_o(\Sigma)$ *contains at least* n+1 *points, and at least* 2^n *if* $\psi_{1|\Sigma}$ *is transversal to* Σ.

2. HOW THIS IS FORMULATED AS A VARIATIONAL PROBLEM.

Let $s : T^*\mathbb{R}^n \to T^*M$ be the canonical universal covering. The pullback $s^*\lambda$ is the Liouville form $\tilde{\lambda}$ of $T^*\mathbb{R}^n$. Let $(\tilde{\psi}_t)$ denote the isotopy of $T^*\mathbb{R}^n$ obtained by lifting

(ψ_t), ant let $\widetilde{\varphi}_i$, $0 \leq i \leq N$, be the lifting of φ_i defined by

$$\widetilde{\varphi}_i = \widetilde{\psi}_{t_{i+1}} \circ \widetilde{\psi}_{t_i}^{-1} \quad .$$

If $\widetilde{f}_i = f_i \circ s$, it is clear that

$$\widetilde{\varphi}_i^* \widetilde{\lambda} - \widetilde{\lambda} = d\widetilde{f}_i \quad .$$

Let $q : T^*\mathbb{R}^n \to \mathbb{R}^n$ and $p : T^*\mathbb{R}^n = \mathbb{R}^n \times (\mathbb{R}^n)^* \to (\mathbb{R}^n)^*$ denote the canonical projections. Clearly,

$$\widetilde{\lambda} = pdq \quad .$$

If $\widetilde{\Sigma} = s^{-1}(\Sigma)$ is the zero section of $T^*\mathbb{R}^n$, define two spaces E and $\widetilde{E}$ and a (universal) covering projection $r : \widetilde{E} \to E$ by

$$\begin{cases} E = \Sigma \times (T^*\mathbb{R}^n)^N \quad , \quad \widetilde{E} = \widetilde{\Sigma} \times (T^*\mathbb{R}^n)^N \\ r(x,y) = (s(x),y) \quad , \quad x \in \widetilde{\Sigma} \quad , \quad y \in (T^*\mathbb{R}^n)^N \quad . \end{cases}$$

Let $z_o : \widetilde{E} \to \widetilde{\Sigma}$ and $u_1, \ldots, u_N : \widetilde{E} \to T^*\mathbb{R}^n$ denote the projections (thus, $a = (z_o(a), (u_i(a))_{1 \leq i \leq N})$ for every $a \in \widetilde{E}$), and let

$$\begin{cases} x_i := q \circ u_i \quad , \quad y_i := p \circ u_i \quad , \\ z_i^- := \widetilde{\varphi}_{i-1} \circ z_{i-1} \quad , \quad z_i := z_i^- + u_i \quad , \\ q_i := q \circ z_i \quad , \quad p_i := p \circ z_i \quad , \\ q_i^- := q \circ z_i^- \quad , \quad p_i^- := p \circ z_i^- \quad . \end{cases}$$

LEMMA 2 .- *Let $f : E \to \mathbb{R}$ be given by $f \circ r = \widetilde{f}$, where*

$$\widetilde{f} = \sum_{0 \leq i \leq N} \widetilde{f}_i \circ z_i + \sum_{1 \leq i \leq N} p_i^- x_i \quad .$$

The critical points of f are precisely those $(\theta, 0) \in \Sigma \times \{0\} \subset E$ such that $\psi_1(\theta) \in \Sigma$. Therefore, they are in one to one correspondance with $\psi_1(\Sigma) \cap \Sigma$.

<u>Proof</u> : $$d\tilde{f} = \sum_{0\leq i\leq N} z_i^*(\tilde{\varphi}_i^*\tilde{\lambda} - \tilde{\lambda}) + d \sum_{1\leq i\leq N} p_i\, x_i =$$

$$= \sum_{0\leq i\leq N} (p_{i+1}^- \, dq_{i+1}^- - p_i \, dq_i) + \sum_{1\leq i\leq N} (p_i \, dx_i + x_i \, dp_i) =$$

$$= p_{N+1}^- \, dq_{N+1}^- + \sum_{1\leq i\leq N} (p_i^- \, dq_i^- + p_i(dx_i - dq_i) + x_i \, dp_i) \quad ,$$

hence in view of the definitions

(1) $$d\tilde{f} = p_{N+1}^- \, dq_{N+1}^- + \sum_{1\leq i\leq N} (x_i \, dp_i - y_i \, dq_i^-) \quad .$$

Therefore, lemma 2 is a consequence of

LEMMA 3 .- *The mapping* $a \mapsto (q_1^-(a), (q_{i+1}^-(a), p_i(a))_{1\leq i\leq N})$ *is a diffeomorphism* $\tilde{E} \to \mathbb{R}^n \times (T^*\mathbb{R}^n)^N$.

<u>Proof</u> : Clearly, $a \mapsto (z_i(a))_{0\leq i\leq N}$ is a diffeomorphism of $\tilde{E}$. Now, by lemma 1 (ii)-(iii), each mapping $z_i(a) \mapsto (q_{i+1}^-(a), p_i(a))$, $0 \leq i \leq N$, is a diffeomorphism. □

3. <u>KEY LEMMAS</u>.

Let us first state a corollary of lemma 3 :

LEMMA 4 .- *Let* $X_i := q_{i+1}^- - q_i^-$, $Y_i := p_i - p_{i-1}$, $1 \leq i \leq N$, *and let* Σ *(resp.* $\tilde{\Sigma}$*) be canonically identified with* M *(resp.* $\mathbb{R}^n$ *). The mappings* $\Phi : E \to E$ *and* $\tilde{\Phi} : \tilde{E} \to \tilde{E}$ *given by* $r \circ \tilde{\Phi} = \Phi \circ r$ *and* $\tilde{\Phi}(a) = (q_1^-(a), (X_i(a), Y_i(a))_{1\leq i\leq N})$ *are diffeomorphisms.* □

The following lemma is crucial :

LEMMA 5 .- *Let* $g : E \to \mathbb{R}$ *and* $\tilde{g} : \tilde{E} \to \mathbb{R}$ *be defined by* $\tilde{g} = g \circ r = \sum_{1\leq i\leq k\leq N} x_k\, y_i$. *The derivative* $D(\tilde{g} - \tilde{f} \circ \tilde{\Phi}^{-1}) = \tilde{E} \to \tilde{E}^*$ *is bounded.*

<u>Proof</u> : Rewrite (1) as

$$d\tilde{f} = (p_{N+1}^- - \sum_{1\leq i\leq N} y_i)\, d(q_1^- + \sum_{1\leq k\leq N} X_k) + \sum_{1\leq i\leq k\leq N} (x_k \, dY_i + y_i \, dX_k) \quad ,$$

hence

(2) $$d(\tilde{f} \circ \tilde{\Phi}^{-1}) = ((p_{N+1}^- - \Sigma\, y_i) \circ \tilde{\Phi}^{-1})\, d(q_o + \Sigma\, x_k) + \sum_{1\leq i\leq k\leq N} (x_k \circ \tilde{\Phi}^{-1} dy_i + y_i \circ \tilde{\Phi}^{-1} dx_k)$$

and

$$d(\tilde{f}\circ\tilde{\Phi}^{-1}-\tilde{g}) = ((\bar{p}_{N+1}-\Sigma\, y_i)\circ\tilde{\Phi}^{-1})\, d(q_o+\Sigma\, x_k) +$$

$$+ \sum_{1\leqslant i\leqslant k\leqslant N} ((x_k - X_k)\circ\tilde{\Phi}^{-1}\, dy_i + (y_i - Y_i)\circ\tilde{\Phi}^{-1}\, dx_k) \quad .$$

Therefore, we just have to prove that $\bar{p}_{N+1}-\Sigma\, y_i$ and $x_i - X_i$, $y_i - Y_i$, $1\leqslant i\leqslant N$, are bounded. Now, from the definition of the $\tilde{\varphi}_i$'s and the fact (ψ_t) has compact support, it is clear that $q\circ\tilde{\varphi}_i - q$ and $p\circ\tilde{\varphi}_i - p$ are bounded for $0\leqslant i\leqslant N$. Since $\bar{p}_{N+1}-\Sigma\, y_i = \sum_{0\leqslant i\leqslant N} (p\circ\tilde{\varphi}_i - p)\circ z_i$, $X_i - x_i = (q\circ\tilde{\varphi}_i - q)\circ z_i$ and $Y_i - y_i = (p\circ\tilde{\varphi}_{i-1} - p)\circ z_{i-1}$, $1\leqslant i\leqslant N$, this yields the desired result. □

LEMMA 6 .- *An intersection point $\psi_1(\theta)\in\Sigma\cap\psi_1(\Sigma)$ is transversal if and only if the critical point $(\theta,0)$ of f is nondegenerate.*

Proof : Let $G:\tilde{E}\to(\mathbb{R}^n)^*\times(\mathbb{R}^n\times(\mathbb{R}^n)^*)^N$ be defined by

$$G(a) = ((\bar{p}_{N+1}-\Sigma\, y_i)(a) \quad , \quad (x_i(a),y_i(a))_{1\leqslant i\leqslant N}) \quad .$$

From (2), it is clear that, given $\tilde{\theta}\in s^{-1}(\theta)$, the critical point $(\tilde{\theta},0)$ of $\tilde{f}$ is non-degenerate iff the derivative $DG(\tilde{\theta},0)$ is an isomorphism, i.e. iff

$\frac{\partial\bar{p}_{N+1}}{\partial q_o}(\tilde{\theta},0) = D(p\circ\tilde{\psi}_1|_{\tilde{\Sigma}})(\tilde{\theta})$ is invertible. □

4. END OF THE PROOF, CONSEQUENCES AND COMMENTS.

Let ∇g denote the gradient of g with respect to the standard flat Riemannian metric on E, and let $E_o := (T^*\mathbb{R}^n)^N$. Since

$$f(\theta,x) = g_o(x) \quad , \quad \theta\in\Sigma \quad , \quad x\in E_o \quad ,$$

$\nabla g:\Sigma\times E_o\to(T\Sigma)\times TE_o$ is of the form

$$\nabla g(\theta,x) = (0,\nabla g_o(x)) \quad ,$$

where ∇g_o denotes the gradient of g_o with respect to the standard Euclidean metric on E_o. Now, ∇g_o is a hyperbolic linear vectorfield and, by lemma 5, $\nabla g - \nabla(f\circ\Phi^{-1})$ is a bounded vector field. Therefore, a (by now) classical "isolating block" argument -see for instance [4] § 11- shows that $f\circ\Phi^{-1}$, hence f, has at least $n+1$ critical points, and at least 2^n if none of them is degenerate. By lemma 2 and lemma 6, this completes the proof of our main result. ■

From now on, we assume that $n = 2k$ for some positive integer k. Let σ be the *standard symplectic form on* $\mathbb{C}^k$, i.e. the 2-form defined by

$$\sigma(z)(v,w) = \mathrm{Im}(v \cdot \bar{w}) \quad , \quad z,v,w \in \mathbb{C}^k \quad ,$$

where $v \cdot \bar{w}$ is the standard Hermitian product and Im denotes the imaginary part. Viewing M as $\mathbb{C}^k/(\mathbb{Z} + i\mathbb{Z})^k$, the *standard symplectic structure* ω of M is the 2-form on M obtained from σ *via* the canonical projection. A standard fact of life ([4], § 12) is the following

PROPOSITION .- *Let* Ω *be the symplectic form on* $M \times M$ *given by* $\Omega = \pi_2^*\omega - \pi_1^*\omega$, *where* $\pi_1, \pi_2 : M \times M \to M$ *denote the projections. There exists a covering projection* $P : T^*M \to M \times M$ *such that* $P^*\sigma = d\lambda$ *and* $P_{|\Sigma}$ *is a diffeomorphism onto the diagonal* ΔM. ■

Here is a straightforward consequence :

COROLLARY 1 .- *Let* (j_t) *be as in the main theorem. If* $P(j_1(M))$ *is the graph of a mapping* φ_1, *then* φ_1 *preserves* ω *and has at least* $2k+1$ *fixed points, and at least* 2^{2k} *if none of them is degenerate.* ■

Let $(\varphi_t)_{0\leq t\leq 1}$ be an isotopy of M, and let $(\dot{\varphi}_t)_{0\leq t\leq 1}$ be its *infinitesimal generator*, defined by

$$\dot{\varphi}_t := \left(\frac{d}{dt}\varphi_t\right) \circ \varphi_t^{-1} \quad , \quad 0 \leq t \leq 1 \quad .$$

If every φ_t preserves ω, then (φ_t) is called an *isotopy of* ω, and the Lie derivative $L_{\dot{\varphi}_t}\omega$ is identically 0 ; in other words, by the homotopy formula, the interior product $\dot{\varphi}_t \lrcorner \omega$ is a closed 1-form for every t. If $\dot{\varphi}_t \lrcorner \omega$ is exact for every t, then (φ_t) is called a *Hamiltonian isotopy of* ω. We can now state

COROLLARY 2 (Conley-Zehnder [9]) .- *For every Hamiltonian isotopy* (φ_t) *of* ω, *the number of fixed points of* φ_1 *is as least* $2k+1$, *and at least* 2^{2k} *if none of them is degenerate.*

<u>Proof</u> : Since P has the unique path lifting property, there exists a unique family $(j_t)_{0\leq t\leq 1}$ of embeddings $M \to T^*M$, depending smoothly on t, such that $j_o(M) = \Sigma$ and $P \circ j_t(x) = (x, \varphi_t(x))$, $0 \leq t \leq 1$, $x \in M$. Now, isotopy (φ_t) is Hamiltonian iff every j_t is exact Lagrangian, hence our result by corollary 1. ■

Remarks : - The hypothesis of corollary 2 is apparently more restrictive than that of corollary 1. It may be easy to determine whether this is a real restriction. For a whole bunch of such topological questions -some of which *are* difficult-, the reader is referred to [8].

- The finite dimensional variational method we have introduced above (see [8] for a less cryptic but more complicated presentation) has several advantages over the more natural infinite dimensional framework of the original proof : first, the integer N has more geometric meaning than the dimension to which the problem was reduced in [9]-[5]. Second, no casuistry with Sobolev spaces is needed in order to prove that f is C^2 -this is the really artificial part of infinite-dimensional methods in the calculus of variations ; in particular, the proof of lemma 6 above is much simpler that its previous analogue. In the third place, this method might have a rather wide range of applications (see [8] for other examples) and thus comfort the reactionary opinion that the calculus of variations in one variable may have little to do with modern infinite dimensional analysis. For example, F. Laudenbach and J.C. Sikorav used our method to extend the main theorem to any compact manifold M [11].

This work was motivated by conversations with Daniel Bennequin, Alain Chenciner, Jürgen Moser and Eduard Zehnder.

REFERENCES

[1] V.I. ARNOLD, *Sur une propriété topologique des applications globalement canoniques de la mécanique classique*, C.R. Acad. Sc. Paris, t. 261 (Nov. 1965), Groupe 1, 3719-3722.

[2] V.I. ARNOLD, Méthodes mathématiques de la mécanique classique, Mir, Moscou, 1976.

[3] R. BOTT, *Lectures on Morse theory, old and new*, Bull. (new series) of the AMS, Vol. 7, n° 2 (sept. 1982), 331-358.

[4] M. CHAPERON, E. ZEHNDER *Quelques résultats globaux en géométrie symplectique*, Séminaire sud-rhodanien de géométrie III : autour du théorème de Poincaré-Birkhoff, Travaux en cours, Hermann, Paris (1984), 51-121.

[5] M. CHAPERON, *Quelques questions de géométrie symplectique [d'après, entre autres, Poincaré, Arnold, Conley et Zehnder]*, Séminaire Bourbaki 1982-83, Astérisque 105-106 (1983), 231-249.

[7] M. CHAPERON, *Une idée du type "géodésiques brisées" pour les système hamiltoniens*, C.R. Acad. Sc. Paris, t. 298, n° 13 (1984), 293-296.

[8] M. CHAPERON, *Questions de géométrie symplectique*, Séminaire Sud-Rhodanien IV, Balaruc (may 1984), to appear in Travaux en cours, Hermann, Paris.

[9] C.C. CONLEY, E. ZEHNDER, *The Birkhoff-Lewis fixed point theorem and a conjecture of V.I. Arnold*, Inv. Math. 73 (1983), 33-49.

[10] J. MILNOR, Morse theory, Annals of Math. Study 51, Princeton, 1963.

[11] F. LAUDENBACH, J.C. SIKORAV, *Persistance d'intersection avec la section nulle au cours d'une isotopie hamiltonienne dans un fibré cotangent*, preprint, Université Paris-Sud, 1985.

AN "A-PRIORI" ESTIMATE FOR OSCILLATORY-EQUATIONS.

R. Dieckerhoff and E. Zehnder

Mathematisches Institut der
Ruhr-Universität Bochum
4630 Bochum
(West-Germany)

1) Results

It is well known that the longtime behaviour of a timedependent non-linear differential equation

$$(1) \qquad \ddot{x} + f(t,x) = o,$$

f being periodic in t can be very intricate. For example, there are equations having unbounded solutions but with infinitely many zeroes and with nearby unbounded solutions having randomly prescribed numbers of zeroes and also periodic solutions, see V.M. Alekseev [8], K. Sitnikov [7] and J. Moser [4]. In contrast to such unboundednes phenomena one may look for conditions on the nonlinearity, in addition to the condition, that

$$(2) \qquad \frac{1}{x} f(t,x) \to \infty \text{ as } |x| \to \infty,$$

which allow to conclude that all the solutions of the equation are bounded. For example every solution of the equation

$$(3) \qquad \ddot{x} + x^3 = p(t),$$

$p(t+1) = p(t)$ being continuous is bounded. This result, prompted by questions of J.E. Littlewood in [2], is due to G.R. Morris [1]. The aim is to extend the result to a more general class of equations for which in particular the timedependence is involved in the nonlinearity in x.

Theorem 1

Every solution x(t) *of the equation*

$$(4) \qquad \ddot{x} + x^{2n+1} + \sum_{j=o}^{2n} x^j p_j(t) = o, \quad n \geq 1,$$

with $p_j(t+1) = p_j(t)$ *and* $p_j \in C^\infty$ *is bounded, i.e. it exists for all* $t \in \mathbb{R}$ *and*

$$\sup_{\mathbb{R}} (|x(t)| + |\dot{x}(t)|) < \infty .$$

We shall not make the bound explicit in terms of the initial conditions $x(o)$ and $\dot{x}(o)$ of the solution. The proof of theorem 1 is strictly 2-dimensional and based on J. Moser's twist theorem similarly to the proof of Morris result. The underlying idea is as follows. By means of transformation theory the equation is, outside of a large disc D in the $x, \dot{x}$-plane, transformed into a Hamiltonian equation having the following property. Following the solutions from the section $t = o$ to the section $t = 1$ defines a map, the time 1 map ϕ of the flow, which is close to a so called twist map in $\mathbb{R}^2 \setminus D$. By means of the twist theorem one finds large invariant curves diffeomorphic to circles and surrounding the origin in the $(x,\dot{x})$-plane. Every such curve is the base of a time periodic and unter the flow invariant cylinder in the phase space $(x,\dot{x},t) \in \mathbb{R}^2 \times \mathbb{R}$, which confines the solutions in its interior and which therefore leads to a bound of these solutions. It turns out that the solutions starting at $t = o$ on the invariant curves are quasi-periodic:

<u>Theorem 2</u>
There is a (large) $\omega^* > o$ such that for every irrational number $\omega > \omega^*$ satisfying $|\omega - \frac{p}{q}| \geq \frac{1}{2}|p|^{-2-\beta}$ for all integers p and $q \neq o$ with some $\beta > o$, there is a quasiperiodic solution of (4) having frequencies $(\omega,1)$; i.e. there is a smooth function $F(\theta_1,\theta_2)$ periodic of period 1, such that $x(t) := F(\theta_1 + \omega t, \theta_2 + t)$ are solutions of the equation.

It is well known that the invariant curves of the twist-theorem lead to an abbundance of periodic solutions. Proceeding as in [1] by applying the Poincaré-Birkhoff fixed point theorem to the annuli bounded by two suitable invariant curves one finds fixed points and periodic points of the time 1 map of the flow. They give rise to forced oscillations and subharmonic solutions of the equation. One knows, moreover, that these invariant curves are in the closure of the set of periodic points. Since in our case, the set of invariant curves has infinite Lebesgue-measure in $\mathbb{R}^2 \setminus D$ we shall conclude:

<u>Theorem 3</u>
For every integer $m \geq 1$ there are infinitely many periodic solutions of (4) having minimal period m. Moreover the closure of the set of subharmonic solutions of (4) is of infinite Lebesgue measure in the phase space $\mathbb{R}^2 \times S^1$.

For related results see [5] and [6]. We should mention, that up to now there is no genuine generalization of the first part of this statement to higher dimensions, i.e. equations $\dot{x} = J\nabla h(t,x)$, $x \in \mathbb{R}^{2n}$, $n \geq 2$. In the special case of equations of the form $\dot{x} = J\nabla h(x)+p(t)$ Berestycki and Bahri [13] found infinitely many forced oscillations using minimax techniques, under suitable assumptions on the function h.

2. Remark about the smoothnes requirements.

It is likely that in the general case (1) the boundednes of the solution is related to an excessive smoothnes requirement in the x-variable. The role of the smoothnes in the t-variable is not clear. In the special case of (4) where the time is not involved in the nonlinear term, the dependence on t is only required to be continuous. In fact one can prove:

Theorem 4

Every solution of $\ddot{x} + x^{2n+1} + p(t) = o$, $n \geq 1$, *with a continuous* $p(t+1) = p(t)$ *is bounded. Moreover the statements of theorem 2 and theorem 3 hold true.*

In contrast, the proof of the general case of theorem 1 requires an excessive amount of derivatives in the t-variable. In fact the number of derivatives in t we need depends on the size of the t-dependent nonlinear term. It will follow from the proof that the statement of theorem 1 holds true for the equation

$$\ddot{x} + x^{2n+1} + \sum_{j=o}^{\ell} x^j p_j(t) = o,$$

$o \leq \ell \leq 2n$, under the following smoothnes assumption which is weaker than in theorem 1: $p_j \in C^{\nu}(S^1)$, where ν is an integer which depends on ℓ as follows:

$$\nu > \frac{1}{n}(\ell + 3) \text{ if } o \leq \ell \leq n+1$$

$$\nu > 1 + \frac{4}{n} + [\log_2 \frac{n}{2n+1-\ell}] \text{ if } n+1 \leq \ell \leq 2n.$$

It is not clear whether the boundednes phenomenon is related to the smoothnes in the t-variable or whether this requirement is a shortcoming of the proof. We remark that the statement of theorem 1 holds true for a more general class of equations as can be seen from its proof.

3. Outline of the proof

Denote by $(x,y) = \psi(\lambda,\theta)$ with $\lambda > o$ the action and angle variables of the time-independent part of (4), which are given by a symplectic map $\psi\colon \mathbb{R}^+ \times S^1 \to \mathbb{R}^2\backslash\{o\}$. In the new coordinates the equation (4) becomes a timedependent Hamiltonian system with Hamiltonian function of the following form:

$$h(\lambda,\theta,t) = \lambda^a + \sum_{j=1}^{2n+1} G_j(\theta,t)\cdot\lambda^{j\alpha} ,$$

where $G_j \in C^\infty(T^2)$, $\alpha = \frac{1}{n+2}$ and $a = (2n+2)\alpha$. By a sequence of finitely many canonical transformations, the angle-dependent part will be made small for large λ, using transformation theory: Denote by $A_{\lambda_o} \subset \mathbb{R}^+ \times T^2$ the annulus $A_{\lambda_o} := \{(\lambda,\theta,t) \mid \lambda \geq \lambda_o \text{ and } (\theta,t) \in T^2\}$. Given $r > o$, denote by $F(r)$ the set of C^∞ functions in $(\lambda,\theta,t) \in \mathbb{R}^+ \times T^2$, which are defined in $\lambda \geq \lambda_o$ for some $\lambda_o > o$ and for which there is a sequence $\lambda_{j\ell k} > o$ such that

$$\sup_{\substack{\lambda \geq \lambda_{j\ell k}\\ (\theta,t) \in T^2}} \lambda^{j-r}|D_\lambda^j\, D_\theta^\ell\, D_t^k\, f(\lambda,\theta,t)| < \infty .$$

Lemma.

Let

$$H = \lambda^a + h_1(\lambda,t) + h_2(\lambda,\theta,t)$$

with $h_1 \in F(c)$ *and* $h_2 \in F(b)$. *Assume* $a > 1$, $b < a$ *and* $c < a$. *Then there is a canonical diffeomorphism* ψ *depending periodically on* t *of the form*

$$\psi : \lambda = \mu + u(\mu,\phi,t) \quad \text{and} \quad \theta = \phi + v(\mu,\phi,t),$$

with $u \in F(1-(a-b))$ *and* $v \in F(-(a-b))$ *such that* $A_{\mu_+} \subset \psi(A_{\mu_o}) \subset A_{\mu_-}$ *for some large* $\mu_- < \mu_o < \mu_+$. *Moreover the transformed Hamiltonian vectorfield* $\psi^*(X_H) = X_{\hat{H}}$ *is of the form:*

$$\hat{H} = \mu^a + \hat{h}_1(\mu,t) + \hat{h}_2(\mu,\phi,t),$$

where $\hat{h}_1 \in F(c_1)$, *with* $c_1 = \max\{c,b\}$, *is given by* $\hat{h}_1(\mu,t) = h_1 + [h_2]$, *and where* $\hat{h}_2 \in F(b_1)$, *with* $b_1 < b$ *defined by*

$$b_1 = \begin{cases} b-(a-b) & \text{if } b \geq 1 \\ b-(a-1) & \text{if } b \leq 1. \end{cases}$$

Finitely many applications of this Lemma gives a Hamiltonian system outside a large disc, for which the flow does exist for time $o \leq t \leq 1$. Moreover, the time 1 map of this flow is, after a suitable further transformation of the form:

$$\phi \quad : \quad \begin{aligned} \theta_1 &= \theta + \lambda + \hat{f}(\lambda,\theta) \\ \lambda_1 &= \lambda \qquad + \hat{g}(\lambda,\theta), \end{aligned}$$

where the analytic functions $\hat{f}$ and $\hat{g}$ satisfy the estimates

$$|D_\mu^r \, D_\theta^s \, \hat{f}(\lambda,.)| \leq \lambda^{-\frac{\varepsilon}{a-1}}$$

$$|D_\mu^r \, D_\theta^s \, \hat{g}(\lambda,.)| \leq \lambda^{-\frac{\varepsilon}{a-1} - \frac{2-a}{a-1}} \quad ,$$

for $\lambda \geq \lambda^*(r,s)$ and with some constant $\varepsilon > (2-a)$. Moreover, coming from an exact Hamiltonian system, the map ϕ has the intersection property, so that in view of the estimates the assumptions of the twist theorem [3] are met. One therefore concludes the required invariant curves, which are, in these coordinates, close to circles λ = const. Finally the application of the Poincaré-Birkhoff fixed point theorem to the iterated maps ϕ^m, $m \geq 1$ in annuli bounded by two suitable invariant curves establishes the required fixed points and periodic points of minimal period m. Details of the outlined proof will appear elsewhere.

References:

[1] G.R. Morris: "A case of boundedness in Littlewood's problem on oscillatory differential equations", Bull. Austr. Math. Soc. 14 (1976), 71-93.

[2] J.E. Littlewood: "Some problems in real and complex analysis" (Heath, Lexington, Mass. 1969).

[3] J. Moser: "On invariant curves of area-preserving mappings of an annulus", Nachr. Akad. Wiss. Göttingen Math.-Phys. Kl. II, (1962), 1-20.

[4] J. Moser: "Stable and Random Motions in Dynamical Systems", Ann. of Math. Studies Number 77, Princeton, N.J. 1973.

[5] P. Hartman: "On boundary value problems for superlinear second oder differential equations", J. Diff. Eq. 26 (1977), 37-53.

[6] H. Jacobowitz: "Periodic solutions of $\ddot{x} + f(t,x) = o$ via the Poincaré-Birkhoff fixed point theorem", J. Diff. Eq. 20 (1976), 37-52.

[7] K. Sitnikov: "Existence of oscillating motions for three-body problem", Dokl. Akad. Nauk, USSR, 133, No. 2, 303-306, (1960).

[8] V.M. Alekseev: "Quasirandom dynamical systems, I,II,III, Math. USSR Sbornik 5, 73-128 (1968); 6, 505-560 (1968); 7, 1-43 (1969).

[9] G.D. Birkhoff: "Proof of Poincaré's Geometric Theorem", Trans. Amer. Math. Soc. 14 (1913) 14-22.

[10] G.D. Birkhoff: "An Extension of Poincaré's Last Geometric Theorem", Acta Math. 47 (1925).

[11] H. Rüssmann: "On the existence of invariant curves of twist mappings of an annulus", Preprint Mainz (1982).

[12] M. Herman: "Démonstration du théorème des courbes Translatées de nombre de rotation de type constant", manuscript, Paris (1981).

[13] A. Bahri, H. Beresticky: "A perturbation Method in Critical Point theory and Applications", Trans. Amer. Math. Soc.

[14] M. Herman: "Introduction à l'étude des courbes invariantes par les diffeomorphismes de l'anneau", Astérisque (1983).

[15] E. Zehnder: "Periodic solutions of Hamiltonian equations", Springer Lecture Notes in Math. 1031 (1983), 172-213.

[16] J. Pöschel: "Integrability of Hamiltonian Systems on Cantor sets", Comm. on Pure and Appl. Math. XXXVI (1982), 653-695.

The work is partially supported by the Stiftung Volkswagenwerk.

On the structure of germs of vector fields in $\mathbb{R}^3$ whose linear part generates rotations

Freddy Dumortier

Summary

Our aim is to study germs of singularities of vector fields in $\mathbb{R}^3$ whose linear part generates a 1-parameter group of rotations.

We describe how under very general conditions the ∞-jet of the vector field can give information as well on the existence of an invariant C^∞ line and invariant C^∞ cones as on the topology of the singularity. In finite codimension the weak-C°-equivalence class (which is the same as the weak-C°-conjugacy class) is revealed to be determined by a finite jet.

The same is true for the C°-equivalence class of germs in normal form.

However the genuine C°-equivalence class is not necessarily determined by a finite jet, even not by the ∞-jet. There exist non-stabilisable 9-jets, unavoidable in generic 60-parameter families of vector fields on 3-manifolds.

§0. Some notations and definitions

E(n) denotes the ring of germs of C^∞ functions in $0 \in \mathbb{R}^n$

V(n) denotes the space of germs of C^∞ vector fields in $0 \in \mathbb{R}^n$ vanishing in 0

Diff(n) denotes the space of germs of C^∞ diffeomorphisms in $0 \in \mathbb{R}^n$ having the origin as a fixed point

The k-jet of an element $f \in E(n)$, $X \in V(n)$ or $g \in \mathrm{Diff}(n)$ is denoted by resp. $j_k(f)(0)$, $j_k(X)(0)$ or $j_k(g)(0)$, for $k \in \mathbb{N} \cup \{\infty\}$.

Definition 1

Let $X, Y \in V(n)$, we say that X and Y are C^r-conjugate ($r \in \mathbb{N} \cup \{\infty\}$) if there exist local representatives $\tilde{X}$ and $\tilde{Y}$ of resp. X and Y defined on the resp. neighbourhoods V_1 and V_2 of 0 and if there exists some C^r diffeomorphism $h : V_1 \to V_2$ (C^0 diffeomorphism = homeomorphism) such that $h^{-1} \circ \tilde{Y}_t \circ h(x) = \tilde{X}_t$ $\forall$ $x \in V_1$ and $\forall$ $t \in \mathbb{R}$ as long as both sides are defined ($\tilde{X}_t$ denotes the flow of $\tilde{X}$).

Definition 2

Let $f, g \in \mathrm{Diff}(n)$, we say that f and g are C^r-conjugate ($r \in \mathbb{N} \cup \{\infty\}$) if there exist local representatives $\tilde{f}$ and $\tilde{g}$ of resp. f and g defined on the resp. neighbourhoods V_1 and V_2 of 0 and if there exists some C^r diffeomorphism $h : V_1 \to V_2$ such that $h^{-1} \circ \tilde{g} \circ h(x) = \tilde{f}(x)$ $\forall x \in V_1$ as long as both sides are defined.

Definition 3

Let $X, Y \in V(n)$, we say that X and Y are C^r-equivalent ($r \in \mathbb{N} \cup \{\infty\}$) if there exist local representatives $\tilde{X}$ and $\tilde{Y}$ of resp. X and Y defined on the resp. neighbourhoods V_1 and V_2 of 0 and if there exists some C^r diffeomorphism $h: V_1 \to V_2$ which maps integral curves of $\tilde{X}$ to integral curves of $\tilde{Y}$ preserving the sense but not necessarily the parametrization; more precisely : if $p \in V_1$ and $\phi_{\tilde{X}}(p,[0,t]) \subset V$ for $t > 0$ then there is some $t' > 0$ such that $h[\phi_{\tilde{X}}(p,[0,t])] = \phi_{\tilde{Y}}(h(p), [0,t'])$ where $\phi_{\tilde{Z}}(q,s) = \tilde{Z}_s(q)$.

Definition 4

A k-jet T_k of a germ of vector field in V(n) is called stabilisable for C^o-equivalence if there exists some $l > k$ and some l-jet T_l with $j_k(T_l)(0) = T_k$ such that for any X and Y in V(n) with $j_l(X)(0) = j_l(Y)(0)$ we have that X and Y are C^o-equivalent.

Definition 5

A k-jet T_k of a germ of diffeomorphism in Diff(n) is called stabilisable for C^o-conjugacy if there exists some $l > k$ and some l-jet with $j_k(T_l)(0) = T_k$ such that for any f and g in Diff(n) with $j_l(f)(0) = j_l(g)(0)$ we have that f and g are C^o-conjugate.

Definition 6

We say that $X \in V(n)$ fulfils a Łojasiewicz inequality if for some, and hence for any, local representative $\tilde{X}$ there exist $k \in \mathbb{N}$ and $c, \delta > 0$ such that for all x with $\| x \| < \delta$: $\| \tilde{X}(x) \| \geq c \| x \|^k$.

Definition 7

Denote $\gamma : \mathbb{R} \to \mathbb{R}^n \times \mathbb{R} : z \to (0,z)$. Let X be a germ in $0 \in \mathbb{R}^n \times \mathbb{R}$ of a C^∞ vector field on $\mathbb{R}^n \times \mathbb{R}$. We say that X is non-flat along $\{0\} \times \mathbb{R}$ if $j_\infty(X \circ \gamma)(0) \neq 0$.
In the other case we say that X is flat along $\{0\} \times \mathbb{R}$.
Let us also recall a definition introduced by Takens in [Ta]

Definition 8

X, $Y \in V(n)$ are weakly-C^o-equivalent if for some (and hence for all) representatives $\tilde{X}$, $\tilde{Y}$ of X and Y there are bounded neighbourhoods U and V of 0 in $\mathbb{R}^n$ and a homeomorphism $h : U \to V$ such that for any $U_1 \subset U$

with $0 \in U_1$ and any $p \in U_1$:

$$L_{\omega,\tilde{Y},h(U_1)}(h(p)) = h(L_{\omega,\tilde{X},U_1}(p))$$

and

$$L_{\alpha,\tilde{Y},h(U_1)}(h(p)) = h(L_{\alpha,\tilde{X},U_1}(p))$$

In this definition

$L_{\omega,\tilde{X},U_1}(p)$ - the ω-limit set of p with respect to U_1 for $p \in U_1$ - is empty if the positive integral curve of X starting at p leaves U_1 and otherwise is the set of those $q \in \bar{U}_1$ for which there is a sequence $t_1, t_2, \ldots \to +\infty$ with $\lim_{i \to \infty} X_{t_i}(p) = q$.

$L_{\alpha,\tilde{X},U_1}(p)$ is defined analogously replacing "positive integral curve" by "negative integral curve" and "$t_1, t_2, \ldots \to \infty$" by "$t_1, t_2, \ldots \to -\infty$".

$L_{\omega,\tilde{Y},h(U_1)}(h(p))$ and $L_{\alpha,\tilde{Y},h(U_1)}(h(p))$ are defined in the same way.

As proven in [Ta], a homeomorphism $h : U \to V$ realizes a weak-C°-equivalence in between X and Y if

1) $h(K_X) = K_Y$

2) $h|K_X$ is an equivalence in between $X|K_X$ and $Y|K_Y$,

where K_X is the set of those points $p \in U$ for which $L_{\omega,X,U}(p)$ or $L_{\alpha,X,U}(p)$ is non-empty. (Same for K_Y)

Remark If such h as here above is a conjugacy in between $X|K_X$ and $Y|K_Y$, we call it a weak-C°-conjugacy.

Another notion which we use in this paper is the notion of topological transversality ; in order to avoid a misunderstanding let us define what we mean by this for a $(n-1)$-C°-disc with respect to a C^1 vector field.

Definition 9

Let X be a C^1 vector field on $\mathbb{R}^n$, $p \in \mathbb{R}^n$ with $X(p) \neq 0$ and $\phi : B_\varepsilon(0) \subset \mathbb{R}^{n-1} \to \mathbb{R}^n$ a C° embedding with $\phi(0) = p$. We say that ϕ or $\phi(B_\varepsilon(0))$ is topologically transverse to X in p if there exists some flow box B for X around p, i.e. $B = S \times]-\varepsilon,\varepsilon[$ with $S \subset \mathbb{R}^{n-1}$ a nbd of 0 and ψ : $(B,(0,0)) \to (\mathbb{R}^n,p)$ a C^1 diffeo. with $\psi_*(\frac{\partial}{\partial x_n}) = X$ and if there exists a $\varepsilon > \delta > 0$ such that $\pi_1 \circ \psi^{-1} \circ \phi : B_\delta(0) \to S$ is a homeomorphism onto its image, where $\pi_1 : S \times]-\varepsilon,\varepsilon[\to S$ denotes the projection on the first factor.

§1. Starting the study of the vector fields $X \in V(3)$, whose linear part generates rotations; changing into the study of an equivalent $\tilde{X}$.

Let X be a C^∞ vector field in $\mathbb{R}^3$ with a zero in the origin, $X(0) = 0$, and suppose that the linear part of X in 0 i.e. $j_1X(0)$ has as eigenvalues 0 and $\pm\, i\lambda$ with $\lambda \neq 0$; this also means that up to a linear change of coordinates the 1-jet is

$$\lambda\ (x\frac{\partial}{\partial y} - y\frac{\partial}{\partial x}) \tag{1}$$

For these singularities the formal normal form theorem (see [Ta] or [Du]) states that up to a C^∞ change of coordinates the ∞-jet can be given an expression which is invariant under all rotations around the z-axis :

$$\text{i.e. } \lambda(1+f(z,x^2+y^2))(x\frac{\partial}{\partial y} - y\frac{\partial}{\partial x}) + g(z,x^2+y^2)(x\frac{\partial}{\partial x} + y\frac{\partial}{\partial y})$$

$$+\ h(z,x^2+y^2)\frac{\partial}{\partial z} \tag{2}$$

with $f(0,0) = 0$, $g(0,0) = 0$, $h(0,0) = 0$, $\frac{\partial h}{\partial z}(0,0) = 0$.

From now on, we suppose that X fulfils a Łojasiewicz inequality.

This is not too severe a restriction since the vector fields which do not satisfy a Łojasiewicz inequality lie in a pro-algebraic set of infinite codimension in V(3) and can hence be avoided in generic n-parameter families of vector fields on 3- manifolds for any n. (see [Du]).

Moreover: if X is analytic with isolated zero then X fulfils a Łojasiewicz inequality ([Ł]).

Such an X satisfying a Łojasiewicz inequality is of course non-flat along the z-axis and a result of [B.D.]

states that in this case there always exists infinitely near the z-axis a C^∞ invariant curve for X.

Hence we may suppose that -up to a C^∞ change of coordinates- the flow of X leaves the z-axis invariant, or in other words that along the z-axis X is tangent to the z-axis. Moreover from [B.D.] we know the existence of C^∞ cones K of finite contact around $\{0\} \times [0,\infty[$ and $\{0\} \times]-\infty,0]$ inside which one of the following situations occurs (changing X to -X if necessary).

(i) the only orbit of X in K tending to 0 is contained in the z-axis and all the other orbits starting in K leave K after a finite amount of time;

(ii) all the orbits of X in K tend to 0 and their closure is a C^∞ curve which has infinite contact with the z-axis.

To proceed further we again write the vector field X as

$$\lambda(1+f(z,x^2+y^2))(x\frac{\partial}{\partial y} - y\frac{\partial}{\partial x}) + g(z,x^2+y^2)(x\frac{\partial}{\partial x} + y\frac{\partial}{\partial y})$$

$$+ h(z,x^2+y^2)\frac{\partial}{\partial z} + Y_1\frac{\partial}{\partial x} + Y_2\frac{\partial}{\partial y} + Y_3\frac{\partial}{\partial z} \qquad (3)$$

with $f(0,0) = g(0,0) = h(0,0) = \frac{\partial h}{\partial z}(0,0) = 0$, $j_\infty(Y_1)(0) = j_\infty(Y_2)(0) = j_\infty(Y_3)(0) = 0$, $Y_1(0,0,z) = 0$ and $Y_2(0,0,z) = 0 \quad \forall z$.

$\lambda(1+f(z,x^2+y^2))(x\frac{\partial}{\partial y} - y\frac{\partial}{\partial x})+g(z,x^2+y^2)(x\frac{\partial}{\partial x} + y\frac{\partial}{\partial y})+h(z,x^2+y^2)\frac{\partial}{\partial z}$ is here any realization of the analogous infinite jet in (2); the existence of such vector field having the given ∞-jet is guaranteed by the Borel extension theorem.

From now on we also ask the germ of the vector field $g(z,r^2)r\frac{\partial}{\partial r} + h(z,r^2)\frac{\partial}{\partial z}$ to fulfil a Łojasiewicz inequality in the plane.

Again it is easy to see that this condition only excludes a set in V(3) which is contained in a pro-algebraic set of infinite codimension.
However, analytic vector fields in $\mathbb{R}^3$ with isolated singularity do not necessarily fulfil this condition.

The vector field

$$\tilde{X} = \frac{2\pi}{\lambda(1+f(z,x^2+y^2))} X$$

also fulfils this extra condition. We write $\tilde{X}$ as

$$2\pi \left(x\frac{\partial}{\partial y} - y\frac{\partial}{\partial x}\right) + \tilde{g}(z,x^2+y^2)\left(x\frac{\partial}{\partial x} + y\frac{\partial}{\partial y}\right)$$

$$+ \tilde{h}(z,x^2+y^2)\frac{\partial}{\partial z} + \tilde{Y}_1\frac{\partial}{\partial x} + \tilde{Y}_2\frac{\partial}{\partial y} + \tilde{Y}_3\frac{\partial}{\partial z} \qquad (4)$$

with $\tilde{g}(0,0) = \tilde{h}(0,0) = \frac{\partial\tilde{h}}{\partial z}(0,0) = j_\infty(\tilde{Y}_i)(0) = \tilde{Y}_1(0,0,z)$ $= \tilde{Y}_2(0,0,z) = 0$ with $i = 1,2,3$ and $\tilde{g}(z,r^2)\, r\frac{\partial}{\partial r} + \tilde{h}(z,r^2)\frac{\partial}{\partial z}$ fulfils a Łojasiewicz inequality.
For the study of invariant sets and C^o-equivalence we may as well work with the new vector field $\tilde{X}$ instead of X since $\tilde{X}$ is C^∞-equivalent to X by means of the identity.

Take now the halfplane $\{y = 0,\ x \geq 0\}$ and consider the Poincaré-mapping (or first return mapping) $P_{\tilde{X}}$ of $\tilde{X}$ with respect to this half plane. ($P_{\tilde{X}}$ is defined on $\{\ y = 0,\ x > 0\ \}$).

If $\tilde{X}$ would be in normal form -which means that $\tilde{Y} = \tilde{Y}_1\frac{\partial}{\partial x} + \tilde{Y}_2\frac{\partial}{\partial y} + \tilde{Y}_3\frac{\partial}{\partial z}$ could be chosen identically zero- this Poincaré-mapping $P_{\tilde{X}}$ would be the time 1-mapping of $\tilde{g}(z,r^2)\, r\,\frac{\partial}{\partial r} + \tilde{h}(z,r^2)\,\frac{\partial}{\partial z}$

$P_{\tilde{X}}$ is a C^∞ diffeomorphism and can be extended to a C^∞ diffeomorphism on $\{\ y = 0,\ x \geq 0\ \}$ which we also call $P_{\tilde{X}}$: this can f.i. be seen by using a cylindrical blowing up :

$\mathbb{R}^+ \times S^1 \times \mathbb{R} \to \mathbb{R}^3$, $(r,\alpha,z) \to (r\cos\alpha,\ r\sin\alpha, z)$.

In these cylinder-coordinates $\tilde{X}$ gets the expression $2\pi\frac{\partial}{\partial\alpha} + \tilde{g}(z,r^2)\, r\frac{\partial}{\partial r} + \tilde{h}(z,r^2)\frac{\partial}{\partial z} + \tilde{Z}(z,r,\alpha)$ and an easy calculation shows that $j_\infty(\tilde{Z})(\alpha,0,0) = 0 \quad \forall\alpha \in S^1$.
By this it also follows that $P_{\tilde{X}}$ is infinitely near the time 1-mapping of $\tilde{g}(z,r^2)\, r\frac{\partial}{\partial r} + \tilde{h}(z,r^2)\frac{\partial}{\partial z}$.
These diffeomorphisms have extensively been studied in [D.R.R.].
Let us recall the essential results, which we will use in the sequel.
To that end we need to describe the blowing-up method in $\mathbb{R}^2$.

§ 2 The blowing-up method for vector fields $X \in V(2)$ and diffeomorphisms $g \in \mathrm{Diff}(2)$

Let $g : (\mathbb{R}^2,0) \circlearrowleft$ be a C^∞ diffeomorphism, resp. X with $X(0) = 0$ a C^∞ vector field in $\mathbb{R}^2$. Let $\phi : S^1 \times \mathbb{R} \to \mathbb{R}^2$, $(\theta,r) \to (r\cos\theta,\ r\sin\theta)$. Then we can define a C^∞ diffeomorphism $\hat{g}$ on $S^1 \times \mathbb{R}$ with $\phi\circ\hat{g} = g\circ\phi$, resp. a C^∞ vector field $\hat{X}$ on $S^1 \times \mathbb{R}$ with $\phi_*(\hat{X}) = X\circ\phi$.

As a matter of fact $\hat{g}$ and $\hat{X}$ are nothing else but g and X written down in polar coordinates.

If $j_k(g-I)(0) = 0$ then $j_k(\hat{g} - I)(u) = 0 \quad \forall u \in S^1 \times \{0\}$.
Moreover for $k = \infty$ it is clear that for each germ of a C^∞ diffeomorphism $h : (S^1 \times \mathbb{R};\ S^1 \times \{0\}) \circlearrowleft$ along $S^1 \times \{0\}$ with $j_\infty(h - I)(u) = 0\ \forall u \in S^1 \times \{0\}$, there exists some germ of a C^∞ diffeomorphism $k: (\mathbb{R}^2,0) \circlearrowleft$ in 0 with $h = \hat{k}$ and $j_\infty(k-I)(0) = 0$

In the vector field-case with $j_k(X)(0) = 0$ and $j_{k+1}(X)(0) \neq 0$ we may gain information by considering $\overline{X}$ with $\hat{X} = r^k \overline{X}$. $\overline{X}$ is also a C^∞ vector field on $S^1 \times \mathbb{R}$. This division does not alter the orbits of $\hat{X}$ nor their sense, but only the parametrization.
Such division is not permitted for diffeomorphisms.

If- in case of a vector field- $\overline{X}$ still has some bad singularities on $S^1 \times \{0\}$ we can repeat this procedure by blowing up once more in these singularities, and so on.

For more information about the blowing up construction we refer to [Du, Du2], but let us give here a rough description.
In order to blow up, instead of using ϕ and dividing by some power of r, we could as well use $\phi' : \{z \mid \| z \| > \frac{1}{2}\} \subset \mathbb{R}^2 \to \mathbb{R}$, $z \to z - \frac{z}{\| z \|}$ and divide by the same power of $(\| z \| - 1)$.
The vector fields we obtain are analytically equivalent, but the second is now defined on some open domain in $\mathbb{R}^2$ and it gets easier to visualize how we can blow up again in some point $z_o \in \{\| z \| = 1\}$: we just use the mapping $T_{z_o} \circ \phi'$ where T_{z_o} denotes the translation $z \to z + z_o$.

As we again end up on an open domain of $\mathbb{R}^2$ we can repeat the construction if necessary.
For simplicity in notation we denote the first blowing up by ϕ_1, the second by ϕ_2 and so on.

After each sequence of n blowing-ups we find some C^∞ vector field $\overline{X}^n$ defined on some domain $U_n \subset \mathbb{R}^2$.
Let us denote $\Gamma_n = (\phi_1 \circ \dots \circ \phi_n)^{-1}(0) \subset U_n$.

Only one of the connected components of $\mathbb{R}^2\setminus\Gamma_n$, call it A_n, has a non-compact closure.
Furthermore $\partial A_n \subset \Gamma_n$ and ∂A_n, which is homeomorphic to S^1 consists of a finite number of regular closed C^∞ arcs meeting transversally.
$(\phi_1\circ \dots \circ\phi_n)|A_n$ is an analytic diffeomorphism sending A_n onto $\mathbb{R}^2\setminus\{0\}$ and there exists a strictly positive function F_n on A_n such that $\hat{X}^n = F_n \cdot \bar{X}^n$ and $\hat{X}^n\,|\,A_n$ is analytically diffeomorphic with $X\,|\,\mathbb{R}^2\setminus\{0\}$ by means of the diffeomorphism $(\phi_1\circ \dots \circ\phi_n)\,|\,A_n$.

§ 3 Nice decomposition of vector fields $X \in V(2)$

In [Du] it has been proven that when $X \in V(2)$ satisfies a Łojasiewicz inequality there exists a finite sequence of blowing-ups $\phi_1\circ\dots\circ\phi_n$ leading to a germ of a vector field $\bar{X}^n$ along ∂A_n such that the singularities of $\bar{X}^n$ on ∂A_n are

1) isolated singularities p which are hyperbolic or semi hyperbolic with the property that $j_\infty(\bar{X}^n\,|\,W^c)(p) \neq 0$ if W^c is a center manifold for $\bar{X}^n$ in p.
2) regular smooth closed curves (or possibly the whole ∂A_n in case $n = 1$) along which $\bar{X}^n$ is normally hyperbolic.

Moreover the position and the properties of the singularities mentioned above only depend on a finite jet of X.

Furthermore it is always possible to find a finite number of C^∞ lines each cutting ∂A_n in one point, having finite or infinite contact with ∂A_n, and (in case this set is not empty) dividing small neighbourhoods of ∂A_n into a finite number of zones which, after blowing down, provide a decomposition of small neighbourhoods of the singularity into hyperbolic (or saddle) sectors, elliptic sectors and parabolic sectors of attracting or expanding type.

The invariant C^∞ lines in the boundary of these sectors blow down to so called <u>characteristic orbits</u> (or <u>characteristic lines</u>) this means orbits (or an orbit together with the singularity) tending to the singularity with a well defined slope for positive or negative time.
The C^∞ lines having finite contact with ∂A_n give rise to characteristic lines of finite type ; this means characteristic lines possessing a C^∞ parametrization $\gamma : [0,\varepsilon[\to \mathbb{R}^2$ with $j_r\gamma(0) \neq 0$ for some $r \in \mathbb{N}$.
The characteristic lines bordering a hyperbolic sector are always of finite type. And as a matter of fact after joining adjacent parabolic sectors and adding a parabolic sector to an elliptic one if it is adjacent to it, the only possibility for characteristic lines not to be of finite type is between two adjacent elliptic sectors.
Characteristic lines of finite type can also be seen as the graph of a C^∞ function in the variable $x^{1/n}$ for some $n \in \mathbb{N}$ and some C^∞ coordinates (x,y).

§ 4 Successive blowing-ups for a diffeomorphism $g \in \text{Diff}(2)$

In order to apply a succession of blowing-ups to a diffeomorphism $g \in \text{Diff}(2)$ which is infinitely near the time 1-mapping X_1 of a $X \in V(2)$ we proceed as follows : if $\phi_1 \circ \dots \circ \phi_n$ is a succession of blowing-ups for X leading to a nice decomposition as described above, we can also consider $\hat{g}^n$ defined on $A_n \cup \partial A_n$ where $\hat{g}^n$ is a germ of a diffeomorphism on $A_n \cup \partial A_n$ along ∂A_n with the property $\phi_1 \circ \dots \circ \phi_n \circ \hat{g}^n = g \circ \phi_1 \circ \dots \circ \phi_n$.

By construction, in each point of ∂A_n, $\hat{g}^n$ has the same ∞-jet as the time 1-mapping $\hat{X}_1^n$ and moreover $\hat{X}^n = F_n \bar{X}^n$ where F_n is the germ along ∂A_n of a nonegative (even positive on A_n) C^∞ function on $A_n \cup \partial A_n$.

In [D.R.R.] it has been proven that under these circumstances around each point of ∂A_n the diffeomorphism $\hat{g}^n$ is C^∞ conjugate to $\hat{X}_1^n$ by means of a conjugacy which is infinitely near the identity along ∂A_n.
As a consequence infinitely near the C^∞ invariant lines for $\hat{X}_1^n$ we will find C^∞ invariant lines of $\hat{g}^n$; the blowing-down of such a C^∞ invariant line is called a characteristic line for g, which is of finite type if the corresponding characteristic line for X has this property.
These characteristic lines for g also induce a decomposition of small neighbourhoods of 0 into "hyperbolic" "elliptic" and "parabolic" sectors for g.
What can we say concerning the topology of g restricted to such a sector ?

Remember that we still consider g to be infinitely near the time 1-mapping of a vector field X which fulfils a Łojasiewicz-inequality and which has at least one characteristic line.

In [D.R.R.] it has been proven that g restricted to a parabolic or an elliptic sector is C°-conjugate to the time 1-mapping of X restricted to the corresponding sector. (We say that $g|S$ is C°-embeddable in a flow). In [Du2] it has been proven that for vector fields of Łojasiewicz-type there exists only one type of elliptic, attracting parabolic or expanding parabolic sector. Although there also exists only one type of hyperbolic sector for vector fields of Łojasiewicz type [Du2], this is no longer the case for diffeomorphisms of Łojasiewicz type as is shown in [Du3].

However concerning hyperbolic sectors for our diffeomorphisms we proved in [D.R.R.] that orbits inside the sector only stay but a finite number of iterates inside sufficiently small neighbourhoods of the origin.

§ 5 C^∞ invariant cones for $X \in V(3)$ whose linear part generates rotations

Let us come back to the vector fields $\tilde{X}$, that we were studying in § 1. As the associated Poincaré-mapping $P_{\tilde{X}}$ is infinitely near the time 1-mapping of the vector field $\tilde{X}_R = \tilde{g}(z,r^2)r\frac{\partial}{\partial r} + \tilde{h}(z,r^2)\frac{\partial}{\partial z}$, fulfilling a Łojasiewicz inequality, we obtain the existence of C^∞ characteristic lines for $P_{\tilde{X}}$. These will give rise to C^∞ invariant cones for $\tilde{X}$ as we can see as follows :

Consider again the cylindrical blowing up $\tilde{X}^c$ of $\tilde{X}$ as considered in § 1 and having the expression

$$2\pi\frac{\partial}{\partial\alpha} + \tilde{g}(z,r^2)r\frac{\partial}{\partial r} + \tilde{h}(z,r^2)\frac{\partial}{\partial z} + \tilde{Z}(z,r,\alpha)$$

with $j_\infty(\tilde{Z})(\alpha,0,0) = 0 \quad \forall\alpha \in S^1$.

Taking $\phi = \phi_1 \circ \ldots \circ \phi_n : A_n \cup \partial A_n \to \mathbb{R}^2$ the blowing-up-mapping used to put $\tilde{X}_R$ in a nice decomposition, we blow up $\tilde{X}^c$ by means of the mapping id $\times\, \phi : S^1 \times (A_n \cup \partial A_n) \to S^1 \times \mathbb{R}^2$ in the neighbourhood of $S^1 \times \{(0,0)\}$. (of course without any division).

Let us denote this blowing up by $\hat{\tilde{X}}^{n}_{c}$.

The Poincaré-mapping of this blown-up vector field with respect to $\{0\} \times (A_n \cup \partial A_n)$ is the 2-dimensional blowing up of $P_{\tilde{X}}$ as constructed in § 4 and which we denoted $\hat{P}^n_{\tilde{X}}$.

A C^∞ characteristic line for $P_{\tilde{X}}$ is obtained as the blowing down of a C^∞ line γ (C^∞ 1-dimensional manifold) invariant under $\hat{P}^n_{\tilde{X}}$ and infinitely near a C^∞ line γ' for $\hat{\tilde{X}}^n_R$ (the 2-dimensional blowing-up of $\tilde{X}_R$ by means of ϕ).

The suspension of γ by means of the orbits of $\hat{\tilde{X}}^{n}_{c}$ defines a (germ of a) C^∞ 2-dimensional invariant manifold $\hat{\Gamma}$, infinitely tangent to $S^1 \times \gamma'$ along $S^1 \times \{(0,0)\}$.

As a matter of fact $S^1 \times \gamma'$ is invariant under the vector field obtained by applying the blowing-up id $\times\, \phi$ to $2\pi\frac{\partial}{\partial\alpha} + \tilde{g}(z,r^2)r\frac{\partial}{\partial r} + \tilde{h}(z,r^2)\frac{\partial}{\partial z}$, while $\hat{\tilde{X}}^n_c$ is infinitely tangent to this vector field along $S^1 \times \{(0,0)\}$ and γ is infintely tangent to γ' in $\{0\} \times \{(0,0)\}$.

Convincing oneself can now most easily be done by using the flow box theorem. [A.R.]

The invariant manifold $\hat{\Gamma}$ can be mapped to $S^1 \times \gamma'$ by means of a diffeomorphism $\hat{g}_\Gamma$ which is infinitely near the identity along $S^1 \times \partial A_n$.

$\hat{g}_\Gamma$ can be blown down to a diffeomorphism g_Γ, infinitely near the identity in the origin and mapping the blow-down Γ of $\hat{\Gamma}$ to the blow-down of $S^1 \times \gamma'$ which is a surface of revolution obtained by rotating a characteristic line around the z-axis.

Therefore let us call Γ a C^∞ cone, meaning that Γ is a diffeomorphic image of a cone of revolution obtained by turning a characteristic line (not lying on the z-axis) around the z-axis.

In many cases -as observed in § 3- this characteristic line is of finite type.

Thanks to the Whitney extension theorem [A.R.] we hence obtain that up to a C^∞ change of coordinates - infinitely near the identity in the origin - $\tilde{X}$ can be given an expression such that, besides the z-axis $\tilde{X}$ leaves invariant a finite number of cones of revolution obtained by rotating characteristic lines around the z-axis.

These characteristic lines are the characteristic lines of the reduced vector field $\tilde{g}(z,r^2)r\frac{\partial}{\partial r} + \tilde{h}(z,r^2)\frac{\partial}{\partial z}$ used to divide small neighbourhoods of the origin into the four types of sectors as described in § 3.

Remark We know that $\tilde{g}(z,r^2)r\frac{\partial}{\partial r} + \tilde{h}(z,r^2)\frac{\partial}{\partial z}$ is not uniquely determined but the cones we obtain will be the same up to a diffeomorphism infinitely near the identity.

§ 6 The different sectors of our specific $X \in V(3)$ in between the invariant C^∞ cones

Let us now study the topology of $\tilde{X}$ restricted to a sector in between two such invariant cones, or inside one cone and containing part of the z-axis in its interior.
For the study of the parabolic and the elliptic sectors, we combine a method due to Takens [Ta] with the results of [D.R.R.].

(i) Let us start with a parabolic sector S of attracting type, having two cones of revolution in its boundary. These cones are obtained by rotating the characteristic lines γ_1 and γ_2 around the z-axis.
Let us denote by S_R the sector of the plane $\{y = 0\}$ in between γ_1 and γ_2.
In [D.R.R.] has been proven that $P_{\tilde{X}} \mid S_R$ is C°-conjugate to the time 1-mapping of the vector field

$$\tilde{X}_R = \tilde{g}(z,r^2)r\frac{\partial}{\partial r} + \tilde{h}(z,r^2)\frac{\partial}{\partial z}.$$

This implies that $\tilde{X} \mid S$ is C°-equivalent to the normal form

$$\tilde{X}_N = 2\pi\ (x\ \frac{\partial}{\partial y} - y\ \frac{\partial}{\partial x}) + \tilde{g}(z,x^2+y^2)(x\frac{\partial}{\partial x}+ y\frac{\partial}{\partial y}) + \tilde{h}(z,x^2+y^2)\frac{\partial}{\partial z}$$

restricted to S.
Like in [Du2] we can find in any small neighbourhood of 0 in S_R a curve C connecting the characteristic lines γ_1 and γ_2 and everywhere transverse to $\tilde{X}_R$. Clearly $S^1 \times C$ is everywhere transverse to $\tilde{X}_N$.
Because of the C°-equivalence, for $\tilde{X} \mid S$ and for $X \mid S$ we also find such C° annulus everywhere topologically transverse to $\tilde{X}$ and X and cutting once all orbits of $X \mid S$ in a sufficiently small neighbourhood.

Two such annuli being homeomorphic one easily proves that not only $X | S$ is C°-conjugate to $X_N | S$ with

$$X_N = \lambda(1+f(z,x^2+y^2))(x\frac{\partial}{\partial y} - y\frac{\partial}{\partial x}) + g(z,x^2+y^2)(x\frac{\partial}{\partial x} + y\frac{\partial}{\partial y})$$

$$+ h(z,x^2+y^2)\frac{\partial}{\partial z}$$ (see § 1) but also that any two such situations $X | S$ and $Y | S$ are C°-conjugate.

<u>Remark</u> Parabolic sectors of expanding type having two cones of revolution in their boundary can be treated in exactly the same way.

(ii) Take now the parabolic sector of attracting type having one cone of revolution in its boundary and containing f.i. the upper half of the z-axis in its interior.

As we did already in § 1 we reconsider the cylindrical blowing up of $\tilde{X}$ which we denoted $\tilde{X}^c$ with

$$\tilde{X}^c = 2\pi\frac{\partial}{\partial\alpha} + \tilde{g}(z,r^2)r\frac{\partial}{\partial r} + \tilde{h}(z,r^2)\frac{\partial}{\partial z} + \tilde{Z}(z,r,\alpha)$$

The blown up of the sector S is a sector $S^c \subset S^1 \times \mathbb{R}^2$, with 2 cones of revolution of resp. γ_1 and γ_2 in its boundary and where again the Poincaré mapping $P_{\tilde{X}^c}$ with respect to $\{0\} \times \mathbb{R}^2$ is C°-conjugate to the time 1-mapping of $\tilde{g}(z,r^2)r\frac{\partial}{\partial r} + \tilde{h}(z,r^2)\frac{\partial}{\partial z}$.

In that way $\tilde{X}^c | S^c$ is C°-equivalent to $2\pi\frac{\partial}{\partial\alpha} + \tilde{g}(z,r^2)r\frac{\partial}{\partial r} + \tilde{h}(z,r^2)\frac{\partial}{\partial z}$ restricted to S^c, and we again find in any small neighbourhood some curve C connecting the characteristic lines γ_1 and γ_2 such that $S^1 \times C$ is everywhere transverse to $2\pi\frac{\partial}{\partial\alpha} + \tilde{g}(z,r^2)r\frac{\partial}{\partial r} + \tilde{h}(z,r^2)\frac{\partial}{\partial z}$.

This annulus $S^1 \times C$ blows down to a disk D which is everywhere tranverse to

$$2\pi(x\frac{\partial}{\partial y} - y\frac{\partial}{\partial x}) + \tilde{g}(z,x^2+y^2)(x\frac{\partial}{\partial x} + y\frac{\partial}{\partial y}) + \tilde{h}(z,x^2+y^2)\frac{\partial}{\partial z}$$

and which cuts every orbit of that vector field laying inside S.

Because of the C°-equivalence we also find a topological disk having the same properties with respect to $X \mid S$.

In that way we again can prove that $X \mid S$ is C°-conjugate to $X_N \mid S$ and also that any two such situations $X \mid S$ and $Y \mid S$ are C°-conjugate.

<u>Remark</u> The same applies in the expanding case.

(iii) For the parabolic sector of attracting type (or expanding type) containing a full neighbourhood of the origin we can perform a same kind of analysis leading to the existence in any small nbd. of 0 of a topological sphere surrounding the origin, topologically transverse to X and cutting each orbit of X except the origin in exactly one point.

This again induces a unique model up to C°-conjugacy.

(iv) Consider now the elliptic sector having two cones of revolution in its boundary. The cones are obtained by rotating the characteristic lines γ_1 and γ_2 around the z-axis. We take $P_{\tilde{X}}$ to be expanding along γ_1 and contracting along γ_2 and take S_R to be the sector in the $\{y = 0\}$-plane in between γ_1 and γ_2 ; $S = S^1 \times S_R$.

Again, by [D.R.R.] $P_{\tilde{X}} \mid S_R$ is C°-conjugate to the time 1-mapping of $\tilde{g}(z,r^2)r\frac{\partial}{\partial r} + \tilde{h}(z,r^2)\frac{\partial}{\partial z}$ restricted to S_R, and hence $\tilde{X} \mid S$ is C°-equivalent to the normal form $\tilde{X}_N \mid S$.

As proven in [Du2], inside S_R we can find some figure (like in the picture) consisting of three arcs α,β,γ which are topologically transverse to the orbits of $\tilde{g}(z,r^2)r\frac{\partial}{\partial r} + \tilde{h}(z,r^2)\frac{\partial}{\partial z}$.

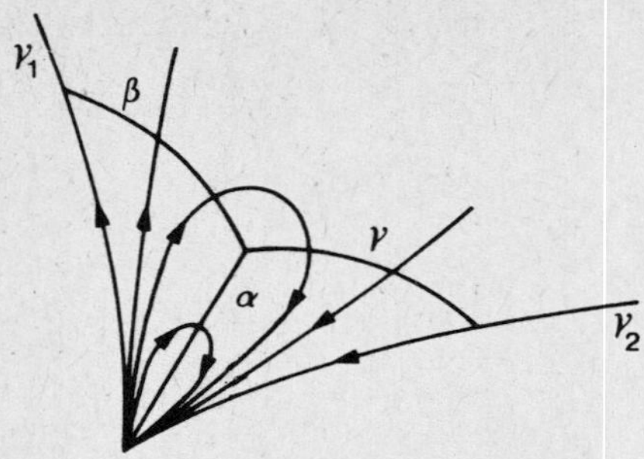

The surfaces $S^1\times\alpha$, $S^1\times\beta$ and $S^1\times\gamma$ are topologically transverse to $\tilde{X}_N$ and induce a subdivision of S into a first region S_1 bounded by $S^1\times\gamma_1$, $S^1\times\alpha$ and $S^1\times\beta$ and a second region S_2 bounded by $S^1\times\gamma_2$, $S^1\times\alpha$ and $S^1\times\gamma$.
As $\tilde{X}\,|\,S$ is C°-equivalent to $\tilde{X}_N\,|\,S$ we can find a same kind of configuration for $\tilde{X}\,|\,S$.
Let us denote $S^1\times\alpha = A$, $S^1\times\beta = B$ and $S^1\times\gamma = C$.
Having a second such vector field $\tilde{Y}\,|\,S$ with associated A', B' and C' we ascertain that any homeomorphism from $A\cup B\cup C \to A'\cup B'\cup C'$ sending A to A', B to B' and C to C' can be extended to a C°-conjugacy in between $\tilde{X}\,|\,S$ and $\tilde{Y}\,|\,S$.
Hence also in this case we see that $\tilde{X}\,|\,S$ is C°-conjugate to a normal form $\tilde{X}_N\,|\,S$ and that we only have but one model of such sector up to C°-conjugacy.

(v) Take now the elliptic sector having one cone of revolution in its boundary and containing f.i. the upper half of the z-axis in its interior.

Like in the corresponding parabolic case we first consider the cylindrical blowing up $\tilde{X}_c$ on the blown-up S_c of the sector S; on S_c we construct $S^1 \times \alpha$, $S^1 \times \beta$ and $S^1 \times \gamma$ as we did in the previous step, but moreover we construct β in a way that $S^1 \times \beta$ blows down to a disk. We then can proceed exactly like before showing that $\tilde{X} \mid S$ in C°-conjugate to the normal form $\tilde{X}_N \mid S$. For such a sector we only have but two models up to C°-conjugacy, one where the vector field on the bordering cone is attracting and one where it is expanding.

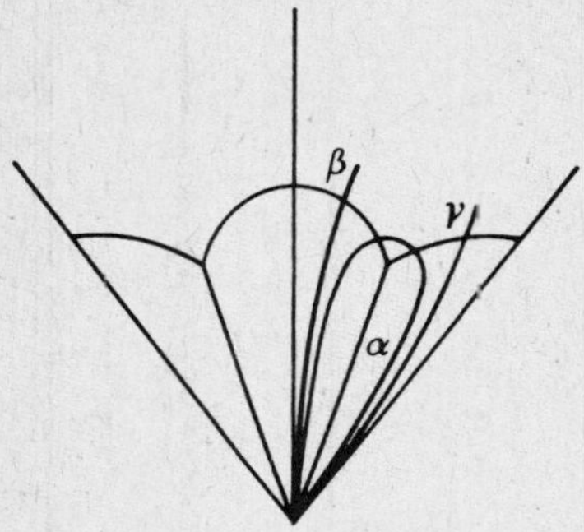

(vi) Finally is left the elliptic sector containing a full neighbourhood of the origin.

Exactly like before we consider the cylindrical blowing up $\tilde{X}_c$ and again construct $S^1 \times \alpha$, $S^1 \times \beta$ and $S^1 \times \gamma$ but in a way that $S^1 \times \beta$ and $S^1 \times \gamma$ both blow down to a disk. We again are left with one model up to C°-conjugacy.

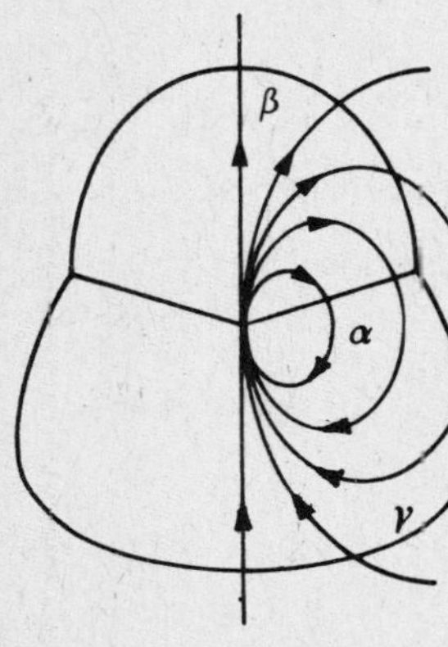

§ 7 A certain property of hyperbolic sectors for our specific $X \in V(3)$

The study of the hyperbolic sectors is much more delicate.

(i) Let us again start with a hyperbolic sector S having two cones of revolution in its boundary.

As we already remarked, in [D.R.R] it has been proven that for any neighbourhood V of 0 there exists a neighbourhood W of 0 with $W \subset V$ such that orbits of $P_{\tilde{X}}$ in $S_R \setminus (\gamma_1 \cup \gamma_2)$ (we say in the interior of S_R) only stay but a finite number of iterates inside $W \cap S_R$.
Considering $S^1 \times W = \{(x,y,z) \mid ((x^2+y^2)^{1/2},z) \in W\}$, we obtain that orbits of $\tilde{X}$ in the interior of S only stay but a finite amount of time inside $S^1 \times W$.

Remark The same analysis applies for a hyperbolic sector having one cone of revolution in its boundary and for a hyperbolic sector containing a full neighbourhood of the origin.

§ 8 Weak-C^o-conjugacy for the $X \in V(3)$ whose linear part generates rotations

As in all our previous constructions we have sufficient freedom on the choice of C, α, β, γ and on the homeomorphisms conjugating X in its different sectors with X_N, we obtain as a result that X is weakly-C^o-conjugate to X_N, and that the weak-C^o-conjugacy class of X only depends on the subdivision into different sectors (the "scheme of sectors"). Moreover, as the subdivision into different sectors only depends on some finite jet of X -as follows from the construction and from [Du]- we see that X is determined by some finite jet up to weak-C^o-conjugacy. This means that for any $X \in V(3)$ with all properties as stated before there always exists some finite number $n \in \mathbb{N}$ such that $\forall\, Y \in V(3)$ with $j_n(X)(0) = j_n(Y)(0)$ we obtain that X is weakly-C^o-conjugate to Y.

§ 9 C°-equivalence for normal forms of the $X \in V(3)$ whose linear part generates rotations

The problem to extend the properties obtained for weak-C°-conjugacy to C°-equivalence is caused by the hyperbolic sectors.

To say more about these hyperbolic sectors we start by detecting the kind of hyperbolic sectors occuring for vector fields in normal form, i.e. for X_N. We begin with

$$\tilde{X}_N = 2\pi(x\frac{\partial}{\partial y} - y\frac{\partial}{\partial x}) + \tilde{g}(z,x^2+y^2)(x\frac{\partial}{\partial x} + y\frac{\partial}{\partial y}) + \tilde{h}(z,x^2+y^2)\frac{\partial}{\partial z}$$

restricted to a hyperbolic sector S having two cones of revolution S^1 "×" γ_1 and S^1 "×" γ_2 in its boundary.

Take $\tilde{X}_N$ to be attracting on $S^1 \times \gamma_1$ and expanding on $S^1 \times \gamma_2$ $P_{\tilde{X}_N}$ is exactly the time 1-mapping of

$$\tilde{X}_R = \tilde{g}(z,r^2)r\frac{\partial}{\partial r} + \tilde{h}(z,r^2)\frac{\partial}{\partial z}.$$

If we now take a second such situation $\tilde{Y}_N \,|\, S'$, then since $P_{\tilde{Y}_N}$ is also the time 1-mapping of $\tilde{Y}_R$ and since $\tilde{X}_R$ and $\tilde{Y}_R$ are of Łojasiewicz-type we obtain that $\tilde{X}_R$ and $\tilde{Y}_R$ are C°-conjugate, as was proven in [Du2].

This conjugacy extends naturally to a C°-conjugacy in between $\tilde{X}_N \,|\, S$ and $\tilde{Y}_N \,|\, S'$ expressed in cylindrical coordinates as $(\alpha, r, z) \to (\alpha, h(r,z))$ where h is the conjugacy in between $\tilde{X}_R$ and $\tilde{Y}_R$.

The freedom on the construction again permits to combine this conjugacy with conjugacies in adjacent sectors.

Exactly the same happens for hyperbolic sectors containing a full neighbourhood of the origin and for hyperbolic sectors having only one cone of revolution in their boundary.

In the last case we distinguish between the cases where the vector field is attracting resp. repelling on the cone of revolution.

To summarize, we have proven that when X_N and Y_N are two vector fields in normal form with a same kind of decomposition in elliptic, parabolic and hyperbolic sectors that then $\tilde{X}_N$ and $\tilde{Y}_N$ and C°-conjugate and a fortiori that X_N and Y_N are C°-equivalent.
Again the equivalence class of such X_N only depends on some finite jet.

§ 10 Hyperbolic sectors for vector fields which are not in normal form, and non-stabilisable jets.

In this part we will try to investigate if something new can happen inside a hyperbolic sector for a vector field

$$\tilde{X} = 2\pi(x\frac{\partial}{\partial y} - y\frac{\partial}{\partial x}) + \tilde{g}(z,x^2+y^2)(x\frac{\partial}{\partial x} + y\frac{\partial}{\partial y}) + \tilde{h}(z,x^2+y^2)\frac{\partial}{\partial z} + \tilde{Y}$$

as in § 1 with $\tilde{Y} \neq 0$
As seen in § 5 we may suppose that the z-axis is invariant and that $\tilde{X}$ has a same kind of decomposition in different sectors as $\tilde{X}_N$ has, even using exactly the same invariant C^∞ cones.
Although $\tilde{X}$ is infinitely near the normal form $\tilde{X}$ we will see that in a hyperbolic sector bounded by 2 cones $\tilde{X}$ does not need to be C°-equivalent with $\tilde{X}_N$ restricted to that same sector.

<u>Remark</u> One can prove that nothing new happens in a hyperbolic sector containing the whole z-axis or half of it in its interior.

As it is not our aim is not to give a full description of the different equivalence classes of hyperbolic sectors that one can encounter, but only to show that something new can happen, we will treat a specific example.

Let $X \in V(3)$ have as 9-jet

$$2\pi(1+\alpha(z,x^2+y^2))(x\frac{\partial}{\partial y} - y\frac{\partial}{\partial x}) + (-z^3(x^2+y^2)+z(x^2+y^2)^2)(x\frac{\partial}{\partial x} + y\frac{\partial}{\partial y})$$

$$+ (z^6-2z^4(x^2+y^2) + z^2(x^2+y^2)^2-(x^2+y^2)^4)\frac{\partial}{\partial z} \qquad (1)$$

where $A(z,u) = \alpha(z,u^2)$ is such that $j_3A(0) = 0$

Because of the Seidenberg-Tarski theorem [S] and the formal normal form theorem, already used in § 1, the germs in V(3) which can be reduced to the vector fields with a 9-jet as in (1) -by means of a C^∞ change of coordinates- form a semi-algebraic set of codimension 60.

As we intend to study X up to C°-equivalence we may as well consider $\tilde{X} = (1+ \alpha(z,x^2+y^2))^{-1}X$ which changes the 9-jet into

$$2\pi(x\frac{\partial}{\partial y} - y\frac{\partial}{\partial x}) + (-z^3(x^2+y^2) + z(x^2 + y^2)^2)(x\frac{\partial}{\partial x} + y\frac{\partial}{\partial y})$$

$$+ (z^6-2z^4(x^2+y^2) + z^2(x^2+y^2)^2 - (x^2+y^2)^4)\frac{\partial}{\partial z} \qquad (2)$$

We may assume (up to a C^∞ change of coordinates) that $j_\infty(\tilde{X})(0)$ is in normal form (see § 1), and because of the presence of the term $z^6 \frac{\partial}{\partial z}$ we moreover may assume, like in § 1, that (after an extra C^∞ change of coordinates) the z-axis is invariant under $\tilde{X}$.

The theorem of Borel guarantees the existence of some $\tilde{X}_N \in V(3)$ in normal form with $\tilde{X} = \tilde{X}_N + \tilde{Z}$ and $j_\infty(\tilde{Z})(0) = 0$.

In [Du3] we prove that for any such $\tilde{X}_N$ it is possible to construct $\tilde{Z}$ in a way that $\tilde{X}$ and $\tilde{X}_N$ are not C°-equivalent.

Let us now give a rough description of this construction. The detailed version is rather lengthy and can be found in [Du3].

As remarked in § 1, $j_\infty(P_{\tilde{X}} - P_{\tilde{X}_N})(0) = 0$ and $P_{\tilde{X}_N}$ is the time 1-mapping of a vector field in V(2) whose 9-jet is

$$(-z^3r^2 + zr^4)r\frac{\partial}{\partial r} + (z^6 - 2z^4r^2 + z^2r^4 - r^8)\frac{\partial}{\partial z} \tag{3}$$

Performing a blowing up like in § 2 we obtain the following singularity scheme after two blowing-ups :

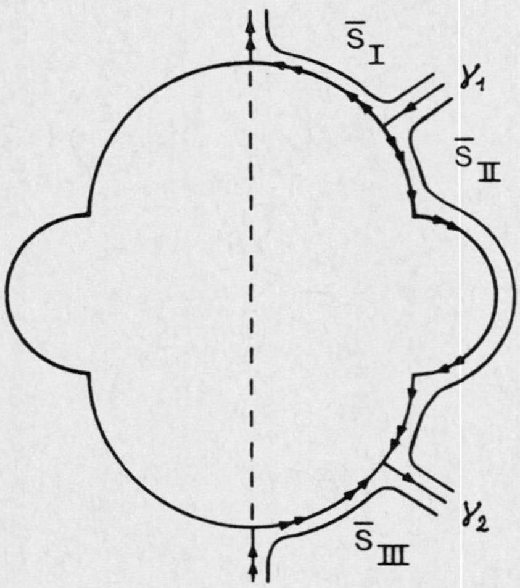

As remarked in § 4, up to a C^∞ change of coordinates (applied to $\tilde{X}$), we may suppose that the blowing downs of resp. γ_1 and γ_2 are characteristic lines (of finite type) for as well $P_{\tilde{X}}$ as for $P_{\tilde{X}_N}$ and that they give rise to invariant C^∞ cones for $\tilde{X}$ and $\tilde{X}_N$. (see § 5).

Let S_I, S_{II}, S_{III} denote the hyperbolic sectors in between these lines (blowing-down of the resp. $\bar{S}_I$, $\bar{S}_{II}$

and $\bar{S}_{III}$ in the picture) and let $\tilde{S}_I$, $\tilde{S}_{II}$ and $\tilde{S}_{III}$ be the corresponding sectors for $\tilde{X}_N$ and $\tilde{X}$.

In [D.R.R] has been proven that $P_{\tilde{X}} \mid S_I$ and $P_{\tilde{X}_N} \mid S_I$ are C^∞-conjugate, just like $P_{\tilde{X}} \mid S_{III}$ and $P_{\tilde{X}_N} \mid S_{III}$ are C^∞-conjugate.

By this it follows that $\tilde{X} \mid \tilde{S}_I$ and $\tilde{X}_N \mid \tilde{S}_I$ are C°-equivalent just like $\tilde{X} \mid \tilde{S}_{III}$ and $\tilde{X}_N \mid \tilde{S}_{III}$ are C°-equivalent.

It is now clear that $\tilde{X}$ and $\tilde{X}_N$ will be C°-equivalent iff $\tilde{X} \mid \tilde{S}_{II}$ and $\tilde{X}_N \mid \tilde{S}_{II}$ are.

In [Du3] we show that $\tilde{X} \mid \tilde{S}_{II}$ and $\tilde{X}_N \mid \tilde{S}_{II}$ are C°-equivalent iff $P_{\tilde{X}} \mid S_{II}$ and $P_{\tilde{X}_N} \mid S_{II}$ are C°-conjugate.

In order to fit our purpose it hence suffices to find $\tilde{Z}$ with $j_\infty(\tilde{Z})(0) = 0$ such that $P_{\tilde{X}_N} \mid S_{II}$ is not C°-conjugate to $P_{\tilde{X}} \mid S_{II}$ (where $\tilde{X} = \tilde{X}_N + \tilde{Z}$).

As $P_{\tilde{X}_N} \mid S_{II}$ is the time 1-mapping of a vector field, by [Du2], $P_{\tilde{X}_N} \mid S_{II}$ is C°-conjugate to $[(x,y) \to (2x, \frac{1}{2} y)] \mid \{x \geq 0, y \geq 0\}$, implying that there must exist invariant C°-foliations F_u and F_s defined in some neighbourhood V of $0 = (0,0) \in \mathbb{R}^2$ and with the following properties :

1) i) Each point of $V \cap S_{II}$ belongs to exactly one leaf of F_u

ii) $\gamma_u \cup \{(0,0)\}$ is a leaf of F_u

iii) each leaf of F_u is connected and cuts γ_s in exactly one point

2) i) Each point of $V \cap S_{II}$ belongs to exactly one leaf of F_s

ii) $\gamma_s \cup \{(0,0)\}$ is a leaf of F_s

iii) each leaf of F_s is connected and cuts γ_u in exactly one point

3) Inside each neighbourhood W of 0 there exists a neighbourhood W' of 0 ($W' \subset V \cap W$) such that
if D is a connected piece of a leaf of F_u cutting γ_s,
if D' is a connected piece of a leaf of F_s cutting γ_u,
if $(X_1)^n(D) \subset W'$ for all $0 \leq n \leq n_1$
if $(X_1)^{-n}(D') \subset W'$ for all $0 \leq n \leq n_2$
then $\# [(X_1)^{n_1}(D) \cap (X_1)^{-n_2}(D')] \leq 1$.

Now $\tilde{Z}$ will be constructed in a way that there will be an obstruction for $P_{\tilde{X}} | S_{II}$ to possess C°-invariant foliations F_u and F_s with the properties as listed above (Especially the third property is revealed to play a central role). To that end we first analyse more closely invariant C^∞ foliations for $P_{\tilde{X}_N} | S_{II}$ by looking at the blowing up $\bar{P}_{\tilde{X}_N}$ (see § 4) of $P_{\tilde{X}_N}$ on $\bar{S}_{II}$.

In the singularity scheme of $\bar{P}_{\tilde{X}_N}$ on $\bar{S}_{II}$ we have singularities p_1, p_2, p_3, p_4 and open arcs γ_1, B, C, D, γ_2 as drawn in the next picture.

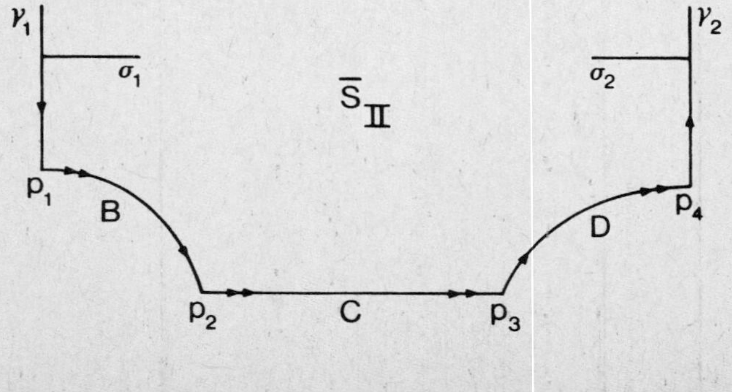

<u>Warning</u> : Double arrows in the picture represent hyperbolic behaviour for the (vector field-) blowing-up of the reduced vector field $\tilde{X}_R$ of $\tilde{X}_N$, but this does not mean that $\bar{P}_{\tilde{X}_N}$ has hyperbolic behaviour there since -see § 4- $\bar{P}_{\tilde{X}_N}$ is the time 1-mapping of a vector field obtained by blowing up $\tilde{X}_R$ without performing the successive divisions by powers of r. Let us denote this blowing up (without division) by $\bar{\bar{X}}_R$. $\bar{P}_{\tilde{X}_N}$ is the time 1-mapping of $\bar{\bar{X}}_R$.
Let σ_1 (resp.σ_2) be a C^∞ segment cutting γ_1 (resp. γ_2) transversally and let F_1 (resp. F_2) be the C^∞ foliation in $\bar{S}_{II}$ obtained by iterating the segment σ_1 (resp. σ_2) by means of the flow of $\bar{\bar{X}}_R$.

In any point b $\in$ B (resp. c $\in$ C or d $\in$ D) we can take a segment cutting B transversally in b (resp. C in c or D in d) and calculate the angle which the leaves of F_1 make with the $\bar{\bar{X}}_R$-orbits along that segment. Especially the asymptotic properties of these angles when approaching B (resp. C or D) reveal to be important.
They f.i. do not depend on the choice of point b (resp c or d) neither on the coordinates in which we perform the calculations.
In terms of any regular parameter r on the segment chosen to cut B transversally in b with r = 0 on B, one can prove that the angle tends to zero as a r^k for some $k \geq 2$ (k depends on the behaviour of $\bar{\bar{X}}_R$ on γ_1).
However in terms of any regular parameter r on the segment chosen to cut C transversally in c (with r = 0 on C) one can prove that the angle tends to zero in a infinitely flat way.
Exactly the same happens for the angles in between F_2 and the $\bar{\bar{X}}_R$-orbits along segments transverse to D and C instead of B and C.

This implies that along a segment cutting C transversally in c, the angle in between F_1 and F_2 will also tend to zero in a flat way.

This will enable us to change $\bar{P}_{\tilde{X}_N}$ in the neighbourhood of c into $\bar{P}_{\tilde{X}}$ - merely by adding a $\tilde{Z}$ to $\tilde{X}_N$ with $j_\infty(\tilde{Z})(0) = 0$ - in a way that the following phenomenon occurs for $\bar{P}_{\tilde{X}}$: if we denote $\tilde{\Sigma}_1 = \cup\{(\bar{\tilde{X}}_R)_t(\tilde{\sigma}_1) \mid t \in]-\frac{1}{3}, \frac{1}{3}[\}$ and $\tilde{\Sigma}_2 = \cup\{(\bar{\tilde{X}}_R)_t(\tilde{\sigma}_2) \mid t \in]-\frac{1}{3}, \frac{1}{3}[\}$ for $\tilde{\sigma}_1$ (resp. $\tilde{\sigma}_2$) a connected piece of σ_1(resp. σ_2) cutting γ_1 (resp. γ_2), then for any neighbourhood W of $\gamma_1 \cup B \cup C \cup D \cup \gamma_2$ it is always possible to find some $W' \subset W, \tilde{\sigma}_1 \subset \sigma_1, \tilde{\sigma}_2 \subset \sigma_2$, $n_1 \in \mathbb{N}$ and $n_2 \in \mathbb{N}$ with $(\bar{P}_{\tilde{X}})^n(\tilde{\Sigma}_1) \subset W' \quad \forall 0 \leq n \leq n_1$, $(\bar{P}_{\tilde{X}})^{-n}(\tilde{\Sigma}_2) \subset W' \quad \forall 0 \leq n \leq n_2$ and such that $(\bar{P}_{\tilde{X}})^{n_1}(\tilde{\Sigma}_1) \cap (\bar{P}_{\tilde{X}})^{-n_2}(\tilde{\Sigma}_2)$ has at least two connected components (see next picture)

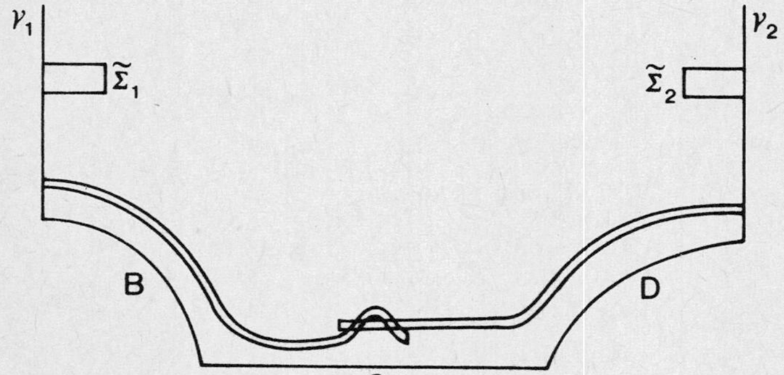

As this phenomenon occurs an infinitely number of times closer and closer to $\gamma_1 \cup B \cup C \cup D \cup \gamma_2$ it is for $\bar{P}_{\tilde{X}}$ impossible to possess invariant C° foliations with all the properties as needed to be C°-conjugate to $\bar{P}_{\tilde{X}_N}$. As a matter of fact invariant C° foliations must in a certain way follow the iterates of the strips $\tilde{\Sigma}_1$ and $\tilde{\Sigma}_2$.

§ 11 Final remarks concerning the topology of vector fields $X \in V(3)$ whose linear part generates rotations

As we just saw in § 10 the topology of germs of vector fields in $\mathbb{R}^3$ whose linear part generates rotations is not necessarily determined by a finite jet up to C°-equivalence; and this not even in finite codimension. The topology is not even necessarily determined by the ∞-jet.

This phenomenon is certainly unavoidable in codimension 60, but presumably already in a codimension ≤ 49. Of course -as we observed in § 8-in finite codimension the topology is finitely determined up to weak-C°-conjugacy. A finite jet hence decides on the weak-C°-conjugacy class, but also on the contact between the different C^∞ invariant cones (see § 5).

In case of germs in normal form and of finite codimension, the topology is finitely determined up to C°-equivalence. A classification of the C°-equivalence classes of normal forms (which are the same as the weak-C°-conjugacy classes) up to a certain codimension is hence reduced to a formal problem.

References

[A.R.] R.Abraham, J. Robbin : Transversal mappings and flows, Benjamin, N.Y. (1967)

[B.D] P. Bonckaert, F. Dumortier : Smooth invariant curves for germs of vector field in $\mathbb{R}^3$ whose linear part generates a rotation.

[Du] F. Dumortier : Singularities of vector fields on the plane J. Diff. Eq. 23 (1977) pp. 53-106

[Du2] F. Dumortier : Singularities of vector fields, Monografias de Matemática nº 32, IMPA, Rio de Janeiro, 1978

[Du3] F. Dumortier : Non-stabilisable jets of diffeomorphisms in $\mathbb{R}^2$ and of vector fields in $\mathbb{R}^3$, to appear

[D.R.] F. Dumortier, R. Roussarie : Germes de difféomorphismes et de champs de vecteurs en classe de différentiabilité finie. Annales de l'institut Fourier. Tome XXXIII, 1, 1983, p 195-267

[D.R.R.] F. Dumortier, P.R. Rodriguez, R. Roussarie Germs of diffeomorphisms in the Plane, Lecture Notes in Mathematics 902, 1981, p 1-197 Springer-Verlag

[Ł] S. Łojasiewics : Ensembles Semi-analytiques, IHES Lecture Notes, 1965

[S] A. Seidenberg : A new decision method for elementary algebra, Ann. of Math 60 (1954) 365-374

[Ta] F. Takens : Singularities of vector fields, Publ. Math. IHES 43 (1974) pp 47-100

Freddy DUMORTIER
Limburgs Universitair Centrum
Universitaire Campus
B-3610 DIEPENBEEK
BELGIUM

FIXED POINT RESULTS FOR SYMPLECTIC MAPS RELATED TO THE ARNOLD - CONJECTURE.

A. Floer and E. Zehnder

Mathematisches Institut der
Ruhr-Universität Bochum
4630 Bochum
(West-Germany)

1) Introduction, history and early results.

Our aim is to present some recent results and open questions concerning the fixed point problem of symplectic maps related to the Arnold-conjecture.

Most of the topological fixed point results originate in the search for periodic solutions in celestial mechanics. Arnold's conjectures belong to the circle of very old questions in celestial mechanics related to the so called Poincaré-Birkhoff fixed point theorem. Recall that in his search for periodic solutions in the restricted 3-body problem, H. Poincaré constructed a section map of an annulus A on the energy surface bound by the direct and retrograde periodic orbits. It lead him to the formulations of the following theorem [19]:

H. Poincaré (1912), G. Birkhoff (1913):
Every area preserving homeomorphism of an annulus $A = S^1 \times [a,b]$ *rotating the two boundaries in opposite direction possesses at least* 2 *fixed points in the interior.*

This result, whose surprising proof by C. Birkhoff, [5] and [6], uses strictly two-dimensional arguments, stands apart from the usual topological fixed point theorems. In fact, it is wrong if the measure preserving condition is dropped, as the following simple example shows. Consider on $(x,y) \in S^1 \times [a,b]$ the map

$$\phi \quad : \quad \begin{aligned} x_1 &= x + \alpha(y) \\ y_1 &= \beta(y) \ , \end{aligned}$$

with $\alpha(a) < o < \alpha(b)$, and with β being strictly monotone and satisfying $\beta(a) = a$, $\beta(b) = b$ and $\beta(y) \neq y$ for $a < y < b$. Then ϕ twists the boundaries of A in opposite direction but obviously has no fixed point.

Nevertheless G. Birkhoff's result gave rise to the topological fixed point theory by S. Lefschetz, which, however, is not applicable to an annulus A since it has Euler-characteristic $\chi(A) = o$, neither is it applicable to the torus $T^2 = \mathbb{R}^2/\mathbb{Z}^2$ to which, at least in the differentiable case, the above result can be reduced,[1],[8]. Recall that for a map f on a compact and oriented space M the Lefschetz-number L(f) is defined by

$$L(f) = \sum_{k=o}^{\infty} (-1)^k \mathrm{Tr}(f_k^*).$$

It is a homotopy invariant such that for $f \simeq id$ it is equal to

$$L(id) = \sum_{k=o}^{\infty} (-1)^k \dim(H_k) = \chi(M).$$

If $L(f) \neq o$ one concludes a fixed point for f. In particular if F_g is the two-dimensional surface of genus g, then

$$\chi(F_g) = 2-2g ,$$

and we conclude that every map f on F_g homotopic to the identity possesses at least one fixed point, provided $g \neq 1$, i.e. provided F_g is not the torus T^2.

In order to describe the impact of the measure preserving condition we consider at first a map f homotopic to the identity on S^2 for which $\chi(S^2) = 2$. It always has at least one fixed point but it may have only one, as the translation $z \to z+1$, $z \in \mathbb{C}$ on the Riemann sphere shows. It is, however, a very striking fact that under the additional assumption that f preserves a regular measure it has ≥ 2 fixed points. This was observed by C. Simon [25] and N. Nikishin [18] in 1974, who proved:

Theorem 1

Every f on S^2 homotopic to the identity and preserving a regular measure has ≥ 2 fixed points.

The following proof is again based on a strictly two dimensional argument used already by C. Loewner in his lectures in the sixties. Assume the fixed point p^* by Lefschetz is the only one. Then f is a homeomorphism of $S^2\setminus\{p^*\} \cong \mathbb{R}^2$ having no fixed point. By Brouwer's translation theorem there is an open set D such that $f^j(D)$ are mutually disjoint. Therefore

$$m(S^2) \geq m\left(\bigcup_{j=o}^{n} f^j(D)\right) = \sum_{j=o}^{n} m(f^j(D)) = (n+1)\, m(D)$$

for every integer n, so that $m(D) = o$ in contradiction to the regularity of the measure m.

As a special case consider a diffeomorphism f on S^2 satisfying $f^*\omega = \omega$ for a volume form ω. By Hopf's theorem it is homotopic to the identity and has, therefore, at least 2 fixed points. In other words it possesses at least as many fixed points as a function on S^2 has critical points.

As already observed the special case of torus $T^2 = \mathbb{R}^2/\mathbb{Z}^2$ is interesting since the Lefschetz theory is not applicable. There are in fact maps close to the identity and measure preserving having no fixed point as the example of the translation shows. Therefore the class of diffeomorphisms on T^2 has to be restricted if it necessarily should have fixed points. We shall consider measure preserving diffeomorphisms ψ which are homologeous to the identity map, hence are, on $\mathbb{R}^2$, represented by

$$\psi : \begin{array}{l} X = x + p(x,y) \\ Y = y + q(x,y) \end{array},$$

with two periodic functions p and q. We shall assume:

(i) ψ is homologeous to id.

(ii) $dX \wedge dY = dx \wedge dy$.

(iii) $\int_{T^2} p\, dx\, dy = o = \int_{T^2} q\, dx\, dy$.

The last condition, equivalent to the invariance of the center of gravity, excludes in particular the translations. For this class of diffeomorphisms on T^2 the following fixed point result is proved in [9]. It was conjectured by V. Arnold in [2] and in fact is a special case of his general conjecture about fixed points of symplectic maps described below.

Theorem 2.
Every diffeomorphism ψ on T^2 satisfying (i)-(iii) has ≥ 3 fixed points. Moreover if all the fixed points are known to be nondegenerate then it has ≥ 4 fixed points.

Here a fixed point p is called nondegenerate, if 1 is not an eigenvalue of the linearized map $d\psi(p)$. This result is obviously not an Euler-characteristic result, it states that the map ψ possesses at least as many fixed points as a function on T^2 has critical points, namely

$$3 = \text{cup long } (T^2) =: CL(T^2)$$
$$4 = \text{sum of the Betti numbers } (T^2) =: SB(T^2).$$

Recall that for a compact space X the integer CL(X) is defined as $CL(X) := 1 + \sup \{k \mid \text{there are classes } \alpha_j \in H^*(X)\setminus\{1\} \text{ such that } \alpha_1 \cup \alpha_2 \cup \dots \cup \alpha_k \neq o\}$, where $H^*(X)$ denotes the Alexander cohomology of X.

In his book [1] Arnold was led to the above result by the following proof, under the additional restriction however that $|\psi\text{-id}|_{C^1}$ is small. It exhibits a direct and one-to-one correspondence between the fixed points of the symplectic map ψ and the critical points of a function on T^2. Namely, following H. Poincaré, [20] 1899, we consider the one-form on T^2:

$$(X-x)(dY + dy) - (Y-y)(dX + dx) = dS(x,y)$$

which, due to assumptions (ii) and (iii) is an exact form, i.e. S is a function on T^2. The fixed points of ψ, $X = x^*$ and $Y = y^*$ correspond in fact to the critical points of S if the two one-forms $(dY + dy)$ and $(dX + dx)$ are linearly independent which is the case if (-1) is not an eigenvalue of $d\psi(x^*,y^*)$, hence in particular if ψ is close to the identity.

The idea of relating fixed points of symplectic maps to critical points of a related function is being used quite frequently in order to establish existence results. For example, A. Weinstein [27] uses it in order to show that a symplectic diffeomorphism of a compact and simply connected manifold M possesses at least as many fixed points as a function on M has critical points, provided the map is sufficiently C^1-close to the identity map on M. For more general results and references we point out J.Moser [17]. So far, however, the method has lead to perturbation results only and one may ask for more global results.

The proof of theorem 2 for a map ψ which is not assumed to be close to the identity uses indeed quite a different idea. The fixed points will also be found as critical points of a function, however, not a function on the finite dimensional manifold T^2, but on the infinite dimensional manifold of contractible loops on T^2.

The crucial observation is the following dynamical interpretation of the assumptions (i)-(iii) on ψ, which will turn the fixed point problem into a problem of dynamical systems and which points to a class of symplectic maps suitable for a fixed point theory on a general symplectic manifold:

Proposition 1.

The following statements for a diffeomorphism ψ on $T^2 = \mathbb{R}^2/\mathbb{Z}^2$ are equivalent:

1) *ψ satisfies the assumptions* (i)-(iii).

2) *$\psi = \phi^1$, the time* 1 *map of the flow ϕ^t of a time dependent exact Hamiltonian vectorfield on T^2:*

$$\frac{d}{dt}\phi^t(x) = J\nabla h(t,\phi^t(x)) \text{ and } \phi^0(x) = x,$$

where $x \in \mathbb{R}^2$ and $h(t,x)$ is periodic in all its variables of period 1.

The matrix $J \in L(\mathbb{R}^2)$ stands for the standard symplectic structure on T^2:

$$J = \begin{pmatrix} 0 & 1 \\ -1 & 0 \end{pmatrix} .$$

The proof of the above proposition for which we refer to [9] is based on a strictly two-dimensional argument using the observation by C.J. Earle and J. Eells that the map ψ being homologeous to the identity actually belongs to the one component of the group of diffeomorphisms on T^2.

In view of the proposition the fixed points of ψ are in one to one correspondence with the 1-periodic solutions of the Hamiltonian equation $\dot{x} = J\nabla h(t,x)$ on T^2, where h is a function periodic in all its variables for period 1. Instead of looking for fixed points of a map we now rather look for periodic solutions of a timedependent Hamiltonian vectorfield! This problem can of course be stated in all dimensions, and the following result, from which Theorem 2 follows, shows that the fixed point theorems discussed so far are not two-dimensional phenomena.

Theorem 3.
Every Hamiltonian vectorfield on $T^{2n} = \mathbb{R}^{2n}/\mathbb{Z}^{2n}$ with its standard symplectic structure $\dot{x} = J\nabla h(t,x)$, $(t,x) \in \mathbb{R} \times \mathbb{R}^{2n}$ and with $h \in C^2$ being periodic in all its variables of period 1 possesses $\geq CL(T^{2n}) = 2n+1$ periodic solutions of period 1. Moreover, $\geq SB(T^{2n}) = 2^{2n}$ if all the 1-periodic solutions are nondegenerate.

Here a 1-periodic solution is called nondegenerate if none of its Floquet-multipliers are equal to 1, or equivalently, if 1 is not an eigenvalue of $d\phi^1(p)$ for the corresponding fixed point $p = \phi^1(p)$. The proof of theorem 3, we refer to [9], is based on a classical variational principle for forced oscillations of Hamiltonian equations; the 1-periodic solutions are found as the critical points of a gradientflow on the space of contractible loops on T^{2n} by using Morse-theory arguments.

In order to formulate the general Arnold conjecture we consider a compact symplectic manifold (M,ω), where ω is the symplectic structure, i.e. a nondegenerate and closed two-form. To a smooth function $H_t : M \to \mathbb{R}$ depending periodically on time t with period 1 one associates the exact Hamiltonian vectorfield X_t on M by

$$\omega(X_t,.) = dH_t(.),$$

using the nondegeneracy of ω. A 1-periodic solution of the Hamiltonian vectorfield is a solution of the equation

$$\frac{d}{dt} x = X_t(x) \text{ on } M$$

satisfying the boundary conditions $x(o) = x(1)$. Arnold conjectures that the Hamiltonian vectorfield on M possesses at least as many 1-periodic solutions as a function on M has critical points hence in view of the Ljusternik-Schnirelman and the Morse-theory this can be stated as

Conjecture of V. Arnold:

$\#$ {*1-periodic solutions*} $\geq CL(M)$
($\geq SB(M)$ *in case all the 1-periodic solutions are nondegenerate*).

We point out, that the conjecture is optimal. In fact in the special case of $M = T^2$ it is easy to construct analytic, exact Hamiltonian vectorfields in any neighborhood of o which possess precisely $CL(T^2) = 3$ periodic solutions of period 1. Since the flow ϕ^t of a Hamiltonian vectorfield consists of symplectic maps i.e. $(\phi^t)^*\omega = \omega$, the 1-periodic solutions give rise to fixed points of the symplectic map $\psi = \phi^1$. The results on periodic solutions of Hamiltonian equations are, therefore, fixed point results for the class of those symplectic maps which can be interpolated by exact Hamiltonian vectorfields. This class, suitable for fixed point theory, has been characterized by A. Banyaga [3]. In order to recall his result let (M,ω) be a compact symplectic manifold and let $\text{Diff}^\infty(M,\omega)$ be the topological group of symplectic C^∞-diffeomorphisms ψ, i.e. $\psi^*\omega = \omega$. With $G \subset \text{Diff}^\infty(M,\omega)$ we denote the identity component, which can be shown to be the identity component by smooth arcs in $\text{Diff}^\infty(M,\omega)$. Then

Theorem 4.

For $\psi \in G$ the following statements are equivalent.

(i) *ψ can be interpolated by an exact Hamiltonian vectorfield on M.*

(ii) *$\psi \in [G,G]$, the commutator subgroup of G.*

(iii) *ψ has vanishing Calabi-invariant.*

For the torus T^2 with the standard symplectic structure the vanishing of the Calabi-invariant for example is equivalent to the invariance of the center of gravity of $\psi \in G$, while in case of S^2 all ψ have vanishing Calabi-invariant, as $H^1(S^2) = o$. As for the definition of this invariant we refer to [3] and [7].

2. The solution of Arnold's conjecture for two-dimensional surfaces F_g via Conley's index theory.

For the two sphere S^2 the conjecture follows immediately from theorem 1. Theorem 2 settles the special case T^2 with the standard symplectic structure. We point out, however, that in dimension two a symplectic form is simply a volume form, and by a well known result of J. Moser [16] any two volumeforms are equivalent, i.e. $f^*\omega_1 = \omega_2$ by a diffeomorphism f, provided they have the same total volume. It is, therefore, sufficient to consider a convenient symplectic structure in the two-dimensional case.

Our aim in the following is to outline a proof of the Arnold-conjecture for a general compact surface F_g having genus $g \geq 1$. For the intricate technical details we refer to [12]. In the following we shall abbreviate

$$F_g = P.$$

Theorem 5.

On a compact two-dimensional surface (P,ω) *having genus* $g \geq 1$ *every exact Hamiltonian vectorfield* $\omega(X_t,.) = dH_t(.)$ *possesses* $\geq CL(P) = 3$ *periodic solutions of period* 1 (resp. $\geq SB(P) = 2+2g$ *in the nondegenerate case*).

We should point out, that this result also follows from Sikorav's theorem mentioned in the next section. In case $g \geq 2$ the existence of at least one solution is, of course, a consequence of the Lefschetz fixed point theory.

a) set up. As convenient symplectic structure we choose one which is adapted to a Kähler metric g on P having nonpositive sectional curvature, such that

$$\omega(X,Y) = g(JX,Y), \qquad X,Y \in T_xP,$$

where J is an endomorphism with $J^2 = -1$ and which is constant under covariant derivatives, $\nabla_X J = o$. In order to formulate the problem in a global setting we denote by ΩP the Hilbertspace of contractible H^1-loops, and denote its tangentbundle by $T(\Omega P)$. With

$$L = \bigcup_{z \in \Omega P} L_z \supset T(\Omega P) = \bigcup_{z \in \Omega P} T_z(\Omega P)$$

we denote the smooth bundle whose fiber L_z consists of L^2 vectorfields over the loop $z \in \Omega P$ with respect to the Riemann metric.
If $H : S^1 \times P \to \mathbb{R}$ is the Hamiltonian function, the vectorfield X_t is defined by $\omega(X_t,.) = dH_t(.)$ and the aim is to find maps $t \to z(t) \in P$ satisfying

$$\frac{d}{dt}z(t) = X_t(z(t)) \text{ and } z(o) = z(1) .$$

Reformulating the problem, the solutions correspond to the zeroes of the section $z \to L_z$ of L given by $\dot{z}(t) - X_t(z(t)) \in T_{z(t)}P$, or equivalently to the zeroes of the section s of L defined by

$$\begin{aligned} s(z)(t) &= J(\dot{z}(t) - X_t(z(t)) \\ &= J\dot{z}(t) + \text{grad } H_t(z(t)), \end{aligned}$$

where $g(\text{grad } H_t(x),.) = dH_t(x)(.)$. Since a continuation argument will be applied we introduce the 1-parameter family of sections

$$s_\lambda(z) = s_o(z) + \lambda \text{ grad } H_t(z) \text{ where } o \leq \lambda \leq 1,$$

with the section $s_o(z) = J\dot{z}$, $z \in \Omega P$, not depending on the Hamiltonian function H.

It is crucial to observe that the section s belongs to a variational functional. In fact, since $\pi_2(P) = o$ we can define the function $a_\lambda : \Omega P \to \mathbb{R}$ by

$$a_\lambda(z) = \int_D \omega + \lambda \int_o^1 H(z(t))dt,$$

D being an immersed two-disc having $z = \partial D$ as boundary. For its derivative one verifies, for $\xi \in T_z\Omega P$:

$$\begin{aligned} da_\lambda(z)(\xi) &= \int_o^1 (\omega(\dot{z}(t),\xi(t)) + \lambda dH_t(z(t))(\xi(t)))dt \\ &= (s_\lambda(z),\xi). \end{aligned}$$

Consequently, since $T_z\Omega P \subset L_z$ is dense, the critical points of the function a_λ on ΩP correspond precisely to the required zeroes of the section s_λ.

In principle, one could search for the equilibrium points of the gradientflow $\frac{d}{ds}z = \nabla a_\lambda(z)$ on ΩP. We shall proceed differently and following the underlying ideas of the proof of the torus case [9] we reduce the problem of finding zeroes of the section s_λ to the problem of finding equilibrium points of a related vectorfield defined on a finite dimensional smooth submanifold $M_\lambda \subset \Omega P$, which will have the right smoothnes and compactnes properties.

b) Reduction to a finite dimensional problem.

Using the fact, that the Holonomy group of the metric is abelian and using spectral properties of the linearization of s_o one can construct, for every integer N, a smooth finite dimensional subbundle $E \subset T(\Omega P) \subset L$ with $E = E_N$ having dimension $\dim(E_z) = (2N+1)\cdot \dim P$, if $z \in \Omega P$.

$$E = \bigcup_{z \in \Omega P} E_z .$$

In L there is an orthogonal splitting $L = E + E^{\perp}$ and a smooth section $\pi \in L(L,L)$ of projections $\pi_z : L_z \to E_z$ and $\pi_z^{\perp} : L_z \to E_z^{\perp}$. Observing that for N sufficiently large the linearization of the map $z \to \pi_z^{\perp} s_\lambda(z) \in E_z^{\perp}$ from ΩP into $E^{\perp}$ is a surjective Fredholm map one finds:

Lemma 1.

Let $o \leq \lambda \leq 1$. *For every* $\rho > o$ *there is an integer* N_o *such that if* $N \geq N_o$ *(recall* $E = E_N$*) the subset of* ΩP:

$$M_\lambda := \{z \in \Omega P \mid \pi_z^{\perp} s_\lambda(z) = o \text{ and } |\dot{z}| < \rho\}$$

is a smooth submanifold of ΩP *having dimension* dimE.

Moreover if $z \in M_\lambda$ *then* $T_z M_\lambda \subset T_z \Omega P$ *and* $L_z = T_z M_\lambda + E_z^{\perp}$ *and* $\pi_z | T_z M_\lambda : T_z M_\lambda \to E_z$ *is an isomorphism.*

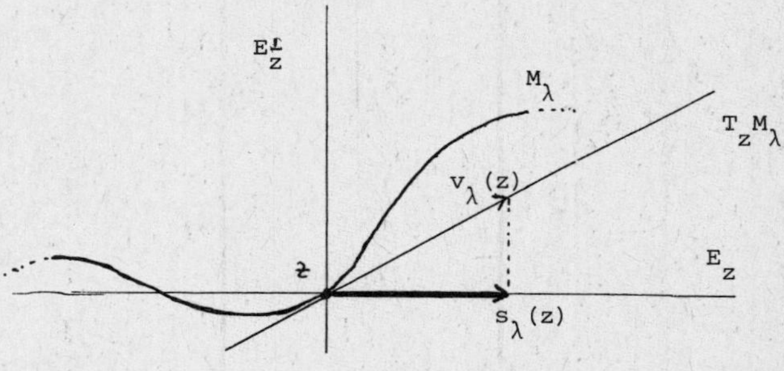

It remains to find the zeroes of $s_\lambda(z)$ for $z \in M_\lambda$ which are the zeroes of the following vectorfield on M_λ:

$$v_\lambda(z) := (\pi_z | T_z M_\lambda)^{-1} \cdot s_\lambda(z) \in T_z M_\lambda ,$$

which still is gradientlike with respect to the old action functional $a_\lambda(z)$, $z \in M_\lambda$.

c. The continuation.

The zeroes of v_λ on M_λ are contained in the set of bounded solutions of the flow of v_λ which we abbreviate by $z\cdot s = \phi_\lambda^s(z)$. Our aim is to show that the set of bounded solutions inherits the topology of the underlying manifold P, which will be proved using Conley's index theory for isolated invariant sets [11]. Using the isometric embedding theorem one concludes that for every $r < \rho$ the set $B_\lambda(r) = \{z \in M_\lambda \mid |\dot{z}| \leq r\}$ is compact. The invariant set of the flow in $B_\lambda(r)$ is defined to be

$$S_\lambda := \{z \in B_\lambda(r) \mid z\cdot s \in B_\lambda(r) \text{ for all } s \in \mathbb{R}\}$$

which a priori may be empty. The following crucial Lemma can be proved estimating the second derivative of the energy $\frac{d^2}{ds^2} E(z\cdot s)$ along a bounded solution and using that the curvature of g is non positive.

Lemma 2.

For r sufficiently large there is an N_o *such that for every* $N \geq N_o$ $(E = E_N)$

$$S_\lambda \subset B_\lambda(\tfrac{1}{2} r), \quad o \leq \lambda \leq 1.$$

Consequently $B_\lambda(r)$ is an isolating neighborhood of the isolated invariant set S_λ. Since the parametrization of the flows is locally a product parametrization, the invariant sets $S_\lambda, o \leq \lambda \leq 1$ are through the lemma related by continuation in the sense of Conley [11] and we shall apply his continuation theory, pointing out D. Salamon [22] for a recent presentation with simplified proofs of this theory. Recall, that for an isolated invariant set S there exists a compact pair (X,Y), the indexpair for S, such that the homotopy type, $[X/Y] = h(S)$ depends only on S and not on the chosen indexpair. It is called the index of S, and is an invariant under continuation; that is, if as in our case the isolated sets S_λ are related by continuation then $h(S_o) = h(S_1)$. Denoting by (X_λ, Y_λ) the index pair of $S_\lambda, o \leq \lambda \leq 1$ there is in particular an isomorphism in the Alexander cohomology

$$j^* : H^*(X_o, Y_o) \to H^*(X_1, Y_1) .$$

This however is not good enough for our purpose. In fact even if $h(S_1) \neq [\emptyset]$, the set S_1 might consist of one single point. We need to know the topology of the invariant set S_1 itself and not only its homotopy index $h(S_1)$. But note that we have an additional map in the surrounding space:

$$\alpha : \Omega P \to P, \text{ defined by } z \to z(o).$$

It gives rise to an additional invariant of the continuation. In fact, if (X_λ, Y_λ) are index pairs of S_λ one defines the product

$$H^*(X_\lambda, Y_\lambda) \otimes H^*(P) \to H^*(X_\lambda, Y_\lambda) \text{ by } (u,\psi) \to u \cup \alpha^*_{X_\lambda}(\psi),$$

where α_{X_λ} is the restriction map $\alpha : X_\lambda \to P$. Using Conley's proof of the continuation theorem one can show that $H^*(X_\lambda, Y_\lambda)$ not only as a ring but also as a $H^*(P)$-modul is independent of λ, in particular:

Lemma 3.

Let S_λ *be related by continuation then, with the above isomorphism* j^*,

$$j^*(u \cup \alpha^*_{X_o}(\psi)) = (j^*u) \cup \alpha^*_{X_1}(\psi),$$

for $u \in H^*(X_o, Y_o)$ *and* $\psi \in H^*(P)$.

The significance of the above invariant lies in the following result which allows to conclude the topology of the invariant set itself from properties of the index pair.

Lemma 4.

Let S *be an isolated invariant set of a continuous flow with index pair* (X,Y). *Assume there is a map* $\alpha : X \to P$. *If there is a class* $u \in H^*(X,Y)$ *such that the map*

$$\psi \to u \cup \alpha^*(\psi) : H^*(P) \to H^*(X,Y)$$

is an isomorphism, then

$$\alpha_S^* : H^*(P) \to H^*(S)$$

is injective, α_S *being the restriction* $\alpha : S \to P$.

Proof.

Define the set $X^- = \{z \in X \mid z \cdot s \in X \text{ for all } s \leq o\}$. By the continuity property of the Alexander-cohomology $H^*(X,X^- \cup Y) = o$, similarly the injection map $S \to X^-$ induces an isomorphism $H^*(S) \cong H^*(X^-)$. Consider now the diagram:

$$\begin{array}{ccccccc}
H^*(X,Y) & \otimes & H^*(X,X^-) & \overset{\cup}{\to} & H^*(X,X^- \cup Y) & = & o \\
\downarrow id & & \downarrow j^* & & \downarrow k^* & & \\
H^*(X,Y) & \otimes & H^*(X) & \overset{\cup}{\to} & H^*(X,Y) & & \\
& & \downarrow i^* & & & & \\
& & H^*(X^-) & & & & \\
& & & \searrow \cong & & & \\
& & & & H^*(S) & &
\end{array}$$

Assume $\alpha_S^*\psi = o$ for some $\psi \in H^*(P)$ then as $\alpha_{X^-} = \alpha_X \circ i$ and since $S \to X^-$ induces an isomorphism, we have $o = \alpha_{X^-}^*\psi = i^* \alpha_X^*\psi$. By the exactness of the short sequence there is a $\phi \in H^*(X,X^-)$ with $j^*\phi = \alpha_X^*\psi$. But then $u \cup \alpha_X^*\psi = k^*(u \cup \phi) = k^*(o) = o$, consequently $\psi = o$ and the map indeed injective. •

In order to apply the lemmata one first observes that the invariant set S_o for $\lambda = o$ consists of the constant loops $z(t) = z(o)$ such that $S_o = P$. Moreover, P can be shown to be a normally hyperbolic invariant manifold of the flow for $v_o(z)$. Consequently there is an index pair (X_o,Y_o) for S_o such that for some $u_o \in H^*(X_o,Y_o)$ the map

$$H^*(P) \to H^*(X_o\ Y_o) \text{ defined by } \psi \to u_o \cup \alpha_{X_o}^*(\psi)$$

is an isomorphism, the so called Thom isomorphism. We therefore conclude from the continuation Lemma 3 and from the previous Lemma 4 that for the required invariant set S_1 the map $\alpha_{S_1} : S_1 \to P$ induces a map

$$\alpha^*_{S_1} : H^*(P) \to H^*(S_1)$$

which is injective, so that indeed the invariant set S_1 inherits the topology of the manifold P. In particualr we have

$$CL(S_1) \geq CL(P) = 3.$$

Since for any continuous and gradientlike flow on a compact set K we can estimate the number of restpoints from below by

$$\# \{\text{rest points}\} \geq CL(K),$$

see [9], we conclude that the vectorfield $v_1(z)$ on $M_1 \subset \Omega P$ has at least 3 zeroes recalling that $CL(P) = 3$. Therefore, the section $s : \Omega P \to L$ has at least 3 zeroes which are the required periodic solutions of the Hamiltonian vectorfield. This finishes the outline of the proof of the first statement.

The second statement will be proved by means of a Morse-theory argument. One first observes

Lemma 5.
A singular point of the vectorfield $v_i(z)$ *on* M_1 *is nondegenerate precisely if it corresponds to a 1-periodic orbit which is nondegenerate in the dynamical sense defined above.*

If all the 1-periodic solutions are nondegenerate there are only finitely many of them. Their corresponding equilibrium points $z_1, \ldots, z_N$ of the flow on S_1 are nondegenerate in M_1 and constitute therefore a Morse-decomposition of S_1. Since the Conley-index of a nondegenerate equilibrium point is the homotopy type of a pointed sphere of the dimension equal to the Morse-index, we find for the Morse-inequalities the equation (see [30]):

$$\sum_{j=1}^{N} t^{m_j} = P(t, X_1, Y_1) + (1+t)\, Q(t),$$

where P is the Poincaré-polynomial for an index pair (X_1, Y_1) for S_1 and where $Q(t)$ is a polynomial having nonnegative integer coefficients. From the continuation argument we know that there is an injective map $H^*(P) \to H^*(X_1, Y_1)$ so that

$$SB(X_1, Y_1) \geq SB(P) = 2+2g.$$

Consequently we conclude from the Morse inequalities that
$N = \#\{\text{equilibrium points}\} \geq 2+2g$ as claimed in theorem 5.

3. Recent results in higher dimensions and Concluding remarks.

On the torus T^{2n} (theorem 3) the required periodic solutions are found as critical points of a classical variational principle for contractible loops on the nice covering space $\mathbb{R}^{2n}$ with its standard constant symplectic structure. Since locally on every compact symplectic manifold one meets the same situation for small contractible loops near the constant loops one can conclude by the same arguments the following remarkable general perturbation-result discovered by A. Weinstein [26].

Theorem 6.
Every timedependent Hamiltonian vectorfield x_t *on a compact symplectic manifold* (M, ω) *possesses* $\geq CL(M)$ *(resp.* $\geq SB(M)$*) periodic solutions of period* 1, *provided*

$$\sup_M |\nabla H_t|$$

is sufficiently small.

Turning now to global results we first recall that for $S^2 = \mathbb{C}P^1$ (theorem 1) the proof was by a strictly two dimensional argument. But the result holds in fact true for any dimension as was recently proved by B. Fortune and A. Weinstein in [13] [14]:

Theorem 7.

A symplectic map on the complex projective space $\mathbb{C}P^n$ (with its standard symplectic structure) which can be interpolated by an exact Hamiltonian vectorfield has $\geq n+1 = CL(\mathbb{C}P^n)$ *fixed points.*

Note that $\pi_2(\mathbb{C}P)^n \neq o$ so that the variational approach outlined above is not applicable directly since the corresponding variation functional is multiple valued. This difficulty is circumvented by considering the Hamiltonian system on $\mathbb{C}P^n$ as the reduction (in the sense of J. Marsden and A. Weinstein [15]) of a Hamiltonian system on $\mathbb{C}^{n+1}$. On this covering space then the periodic solutions are again critical points of the classical variational functional, and are found by minimax techniques based on relative index theories in an infinite dimensional setting such as they are presented in [4].

Finally, J. Sikorav [23], [24] proved the Arnold conjecture for compact Kähler manifolds which have constant negative holomorphic curvature, i.e. for compact quotients of the hyperbolic space $\mathbb{H}^n_{\mathbb{C}}$. In fact he proved the following more general result, from which in particular also theorem 5 follows. Here $\langle X,Y\rangle = g(X,Y)$ stands for the Riemannian metric.

Theorem 8.

The Arnold conjecture holds true for compact symplectic manifolds (M,ω) *having the following properties:* (i) $\omega(X,Y) = \langle JX,Y\rangle$ *with* J *an almost complex structure and an isometry* (ii) *the sectional curvature is* $\leq o$ *and* (iii) *there is a constant* $c > o$ *such that*

$$\langle Jd\,\exp_X(Y),\ d\,\exp_X(JY)\rangle \geq c\,|Y|^2$$

for every $X,Y \in T_pM$ *and every* $p \in M$.

The proof uses the underlying ideas of [9]. In [12] a similar theorem is proved with the above condition (iii) replaced by the condition: J is integrable and the Holonomy group of the Levi-Civitá connection is abelian.

We conclude with some general remarks. The fixed point problems for symplectic maps lead, as we have seen, to a variational functional on the infinite dimensional loop space over the symplectic manifold, which is interesting in itself from an analytical point of view. Namely it is bounded neither from below now from above due to the symplectic structure. This is in sharp contrast to the variational problem of the geometric problem of closed geodesics on a Riemannian manifold which is coercive. That such degenerate variational problems can be used effectively for existence proof was demonstrated by P. Rabinowitz in his seminal paper [21] on periodic solutions of

Hamiltonian equations on $\mathbb{R}^{2n}$. We point out that in the Morse theory for closed geodesics each cohomology class in the loop space is represented in the index of some critical point. This is because the functional is bounded from below so that one has an attractor situation. In the free Hamiltonian case this is obviously not true, only the cohomology of the underlying manifold itself has to be represented as the results above show. Actually the hyperbolic structure of the variational functional for the symplectic form helps in showing that the set of bounded solutions of the gradient flow contains the topology of the manifold. Aiming at a proof of the general case with $\pi_2(M) = o$ it seems however more advantageous to apply the minimax techniques by Rabinowitz to get at the critical points directly rather than first to determine the whole set of bounded solutions of the gradient flow.

Periodic points rather than fixed points of symplectic maps on manifolds have so far not been studied extensively. They correspond to the subharmonic solutions of the corresponding Hamiltonian equations. For global results in the special case of the torus T^{2n} we refer to [10]. Locally in a neighborhood of a generic elliptic fixed point, the closure of the periodic points is a set of positive measure, as is well known, see, for example [28]. It would therefore be interesting to find additional conditions guaranteeing a stable fixed point.

References:

[1] Arnold, V.I.: Mathematical methods of classical mechanics. (Appendix 9) Berlin-Heidelberg, New York: Springer 1978.

[2] Arnold, V.I.: Proceedings of symposia in pure mathematics. Vol. XXVIII AMS, p.66, 1976.

[3] Banyagâ, A.: Sur la structure du groupe des difféomorphismes qui préservent une forme symplectique. Comment. Math. Helvetici 53 (1978), 174-227.

[4] H. Beresticki, J.-M. Lasry, G. Mancini, B. Ruf: Existence of Multiple Periodic Orbits on Star-Shaped Hamiltonian Surfaces", to appear 1984.

[5] Birkhoff, G.D.: Proof of Poincaré's Geometric Theorem. Trans.Amer.Math.Soc. 14, (1913), 14-22.

[6] Birkhoff, G.D.: An Extension of Poincaré's Last Geoemtric theorem. Acta Math. 47 (1925)

[7] Chaperon, M.: Quelques Questions de Géométrie symplectique. Séminaire Bourbaki 1982/83, n° 610, to appear in Astérisque.

[8] M. Chaperon, E. Zehnder: Quelques résultats globaux en géométrie symplectique. In: Géométrie symplectique et de contact: antour du théorème de Poincaré - Birkhoff, Hermann, Paris 1984, 51-122.

[9] Conley, C., Zehnder, E.: The Birkhoff-Lewis Fixed point theorem and a Conjecture of V.I. Arnold, Invent. math. 73 (1983), 33-49.

[10] Conley, C., Zehnder, E.: Subharmonic solutions and Morse-theory", Physica 124A (1984), 649-658.

[11] Conley, C.: Isolated invariant sets and the Morse-index. CBMS, Regional Conf. Series in Math. Vol. 38 (1978).

[12] A. Floer: Proof of the conjecture for surfaces and generalizations for certain Kähler manifolds. PhD-Thesis RUB, Bochum 1984.

[13] B. Fortune, A. Weinstein: A symplectic fixed point theorem for complex projective spaces. Preprint, Berkeley 1984.

[14] B. Fortune, PhD. thesis, Univ. of Cal., Berkeley, 1984.

[15] J. Marsden, A. Weinstein: Reduction of symplectic manifolds with symmetry, Reports on Math. Phys. 5(1974), 121-130.

[16] Moser, J.: On the volume elements on a manifold. Transactions Amer. Math. Soc. 120 (1965), 286-294.

[17] Moser, J.: A fixed point theorem in symplectic geometry. Acta Math. 141 (1978), 17-34.

[18] Nikishin, N. Fixed points of diffeomorphisms on the twosphere that preserve area. Funkcional Anal. i Prelozen 8 (1974), 84-85.

[19] Poincaré, H.: Sur un théorème de Géométrie. Rend. Circolo Mat. Palermo 33 (1912), 375-407.

[20] H. Poincaré: "Les méthodes nouvelles de la mécanique célèste", Gauthiers Villars Paris 1899, tome III, p. 214.

[21] Rabinowitz, P.: Periodic solutions of Hamiltonian systems. Comm. Pure Appl. Math. 31 (1978), 157-184.

[22] Salamon, D.: Connected simple systems and the Conley Index of Isolated invariant sets. MRC Technical summary Report 2753, 1984, Madison.

[23] J. Sikorav: Points fixes d'une symplectomorphisme homologue à l'identité. To appear in C.R. Acad. Sc. Paris, 1984.

[24] J. Sikorav: Points fixes d'une application symplectique homologue à l'identité. PhD Thesis Université de Paris-Sud, 1984.

[25] Simon, C.P.: A bound for the Fixed point Index of an Areapreserving map with Applications to Mechanics. Invent. math. 26 (1974), 187-200.

[26] Weinstein, A.: C^0 perturbation theorems for symplectic fixed points and Lagrangian intersections. Lecture Notes of the AMS summer Institute on nonlinear Functional Analysis and Applications, Berkeley 1983.

[27] Weinstein, A.:Lectures on symplectic manifolds. CBMS Regional Conference series in math. Vol. 29 (1977).

[28] E. Zehnder: Periodic solutions of Hamiltonian equations. Springer Lecture Notes in Math. 1031 (1983), 172-213.

[29] Birkhoff, G.D.: The restricted problem of three bodies. Rend. Circolo Mat. Palermo 39 (1915), 265-334.

[30] C. Conley and E. Zehnder: Morse-type Index theory for Flows and Periodic Solutions for Hamiltonian equations. Comm. on Pure and Appl. Math. Vol. XXXVII (1984), 207-253.

The work is partially supported by the Stiftung Volkswagenwerk.

TOPOLOGICAL INVARIANTS AS NUMBERS OF TRANSLATION

by

J. PALIS and R. ROUSSARIE*

Introduction

In this paper, we interpret as rotation numbers (more precisely, numbers of translation) two topological invariants which appear in [1]. The first one is part of the results of [1] on <u>saddle-node bifurcations</u>. Recall that a one-parameter family F_μ of C^∞ diffeomorphisms of the line presents at $\mu = 0$ a saddle-node bifurcation if $F_o(0) = 0$, $\frac{d}{dx} F_o(0) = 1$, $\frac{d^2}{dx^2} F_o(0) \neq 0$ and $\frac{d}{d\mu} F_\mu(0) > 0$. Under these conditions it is known that F_o imbeds in a unique C^∞ vector field X_o. The result of [1] that we want to interpret is the following: if $\bar{F}_\mu$ is a second saddle-node bifurcation, in $\mu = 0$, and $H(\mu,x) = (\varphi(\mu), h_\mu(x))$ is a topological conjugacy between F_μ and $\bar{F}_\mu$ (h_μ a family of homeomorphisms, $\varphi(t)$ a reparametrization, and for $\forall\, \mu$: $h_\mu \circ F_\mu = \bar{F}_{\varphi(\mu)} \circ h_\mu$), then h_o is <u>necessarily</u> a conjugacy between $X_o(t)$ and $\bar{X}_o(t)$. (For a vector field Z, we denote by $Z(t)$ its flow; here, $\bar{X}_o$ is the unique vector field in which $\bar{F}_o$ imbeds). That is, the time $t(x,y)$ to go from x to y, $x,y < 0$, by $X_o(t)$ is a topological invariant of the family F_μ.

The second topological invariant is linked to the existence of "quasi-transversal" heteroclinic orbits of diffeomorphisms on a 2-dimensional manifold M^2: suppose that p, q are two saddle points of a diffeomorphism F on M^2, such that the unstable manifold $W^u(p)$ and the stable manifold $W^s(q)$ have some quasi-transversal

---

* This work was advanced principally during a visit of the second author to IMPA, Rio de Janeiro, in October-December of 1983.

(i.e., quadratic) contact point r; let λ and μ the stable and unstable eigenvalues in p and q respectively; then the ratio $\frac{\mathrm{Log}|\lambda|}{\mathrm{Log}|\mu|}$ is a topological invariant of conjugacy. This means, it is invariant by every conjugacy of F with some other diffeomorphism $\bar{F}$, defined on a neighborhood of the saddle connection from p to q (that is, the segment made by the part of $W^u(p)$ between p and r and the part of $W^s(q)$ between r and q).

In the following we want to interpret these topological invariants as <u>numbers of translation</u>. In fact, we need a more general notion of number of translation than the usual one which is defined for homeomorphisms of $\mathbb{R}$ commuting with the translation by one. This generalized notion of "<u>asymptotic number of translation</u>" will be defined and studied in part I.

To apply this notions to the two above mentioned questions, we associate to them a <u>class of geometrical objects</u> ("pair of adapted foliations" in the first case, "adapted system of foliations and segment" in the second case) which allows us to introduce asymptotic numbers of translation in a natural way. Finally, these numbers will be invariant when we change the geometrical object in its class, so that they depend only on the given problem. Natural here means that this class of objects is functorial for topological conjugacies.

For the first problem, we give the result in a more general framework by considering "<u>regular</u>" <u>families</u> of topological flows $X_\mu(t)$, which will be a direct topological generalization of saddle-node bifurcation for the $\mu \geq 0$. (See the precise definition at the beginning of part II). The result, given with more precision in Theorem 7 of part II, is the following:

<u>Theorem</u>: <u>Let</u> X_μ, $\bar{X}_\mu$ <u>be two regular families of flows</u>. <u>Then if</u> $H = (h_\mu, \varphi(\mu))$ <u>is a topological conjugacy of</u> $F_\mu = X_\mu(1)$ <u>with</u> $\bar{F}_\mu = \bar{X}_\mu(1)$, <u>the homeomorphism</u> h_o <u>is a conjugacy between</u> $X_o(t)$ <u>and</u> $\bar{X}_o(t)$.

It is another part of the results of [1] that every C^∞ saddle-node bifurcation of diffeomorphisms F_μ may be imbedded in a regular family of flows $X_\mu(t)$ such that $X_o(t)$ is the unique C^∞ flow in which F_o imbeds (F_μ imbeds in $X_\mu(t)$ means that $F_\mu = X_\mu(1)$). So the theorem implies the result mentioned above for the C^∞ saddle-node bifurcations of diffeomorphisms.

The second problem is treated in part III. In this part, essentially, we interpret the formula $\dfrac{\operatorname{Log}\mu}{\operatorname{Log}\lambda} = -\lim\limits_{i} \dfrac{m_i}{n_i}$ of [1] as a way to calculate an asymptotic number of translation.

I. Asymptotic number of translation.

Let $\mathbb{R}$ be the real axis with variable u, and $T: \mathbb{R} \to \mathbb{R}$ the translation $T(u) = u+1$. We denote by Homeo_T the group of the T-commuting homeomorphisms: $\text{Homeo}_T = \{g: \mathbb{R} \to \mathbb{R} \mid g$ is an homeomorphism such that $g \circ T = T \circ g\}$. This group may be seen as the universal covering of the group of homeomorphisms of the circle $T^1 = \mathbb{R}/\mathbb{Z}$. We recall that for each $f \in \text{Homeo}_T$, we can define a number of translation $\rho(f) = \lim\limits_{n\to\infty} \dfrac{f^n(u)}{n}$. This limit exists and is independent of $u \in \mathbb{R}$; $\rho(f)$ modulo $\mathbb{Z}$ is the rotation number of the homeomorphism induced by f on T^1. In this paper we need more general homeomorphisms defined as in the following:

Definition (quasi-T-commuting homeomorphisms).

$\widetilde{\text{Homeo}}_T = \{g$ homeomorphism of $\mathbb{R} \mid g = G + \psi$ for $G \in \text{Homeo}_T$ and $\psi(u) \to 0$ for $u \to +\infty\}$.

Lemma 1. For $g \in \widetilde{\text{Homeo}}_T$, the decomposition $g = G+\psi$ is unique.

Proof: Suppose that $g = G_1+\psi_1 = G_2+\psi_2$. Then $G_1-G_2 = \psi_2-\psi_1$ is T-periodic and tends to 0 as $u \to +\infty$; so it is identically equal to zero.

Definition. The preceeding lemma allows us to define a retraction

$R: \widetilde{\text{Homeo}}_T \to \text{Homeo}_T$ by $R(g) = G$, and also the <u>asymptotic number of translation</u> $\rho(g)$ by: $\rho(g) = \rho(R(g))$ for every $g \in \widetilde{\text{Homeo}}_T$.

<u>Lemma 2</u>. Let $g \in \widetilde{\text{Homeo}}_T$. Then $\rho(g) = \lim_{n\to\infty} \frac{g^n(u)}{n}$. (So, the usual formula is still valid).

<u>Proof</u>: Let $g = G+\psi$ be the decomposition considered above. Let $u_o \in \mathbb{R}$. We want to show that $\rho(g) = \lim_n \frac{g^n(u_o)}{n}$. For each $\varepsilon > 0$, we define $G_{-\varepsilon}(u) = G(u)-\varepsilon$ and $G_{+\varepsilon}(u) = G(u)+\varepsilon$. There exists $u_1(\varepsilon)$ such that for $u \geq u_1(\varepsilon)$:

$$G_{-\varepsilon}(u) \leq g(u) \leq G_\varepsilon(u). \tag{1}$$

Clearly, $g(u) \to +\infty$ as $u \to +\infty$, and g is increasing. Also, we may assume that, for each $u \in \mathbb{R}$, $g^n(u) \to +\infty$ as $n \to +\infty$, and $g^n(u)$ is increasing with n.

Let N be such that $g^N(u_o) = u_N > u_1$.

We are going to show by recurrence, that:

$$\text{for all } n \in \mathbb{N}, \quad G^n_{-\varepsilon}(u_N) \leq g^n(u_N) \leq G^n_\varepsilon(u_N). \tag{2}$$

This inequality is trivial for $n = 0$. Now the recurrence follows from the following observation:

if α, β, γ are such that $\beta > u_1$ and $\alpha \leq \beta \leq \gamma$

then $G_{-\varepsilon}(\alpha) \leq G_{-\varepsilon}(\beta) \leq g(\beta) \leq G_\varepsilon(\beta) \leq G_\varepsilon(\gamma)$.

This follows from (1) and the fact that $G_{-\varepsilon}, G_\varepsilon$ are increasing. So, $G_{-\varepsilon}(\alpha) \leq g(\beta) \leq G_\varepsilon(\gamma)$.

Now, using that $\overline{\lim} \frac{g^n(u_N)}{n} = \overline{\lim} \frac{g^n(u_o)}{n}$ and $\underline{\lim} \frac{g^n(u_N)}{n} = \underline{\lim} \frac{g^n(u_o)}{n}$, the inequalities in (2) imply:

$$\rho(G_{-\varepsilon}) \leq \underline{\lim} \frac{g^n(u_o)}{n} \leq \overline{\lim} \frac{g^n(u_o)}{n} \leq \rho(G_\varepsilon). \tag{3}$$

But, the (classical) number of translation $\rho(G)$ depends continuously of $G \in \text{Homeo}_T$ (for the uniform topology). So, if $\varepsilon \to 0$

in the formula (3), we see that the sequence $\{\frac{g^n(u_o)}{n}\}$ converges and

$$\lim_{n\to\infty} \frac{g^n(u_o)}{n} = \rho(G) = \rho(g).$$

Lemma 3. $\widetilde{\mathrm{Homeo}}_T$ is a group and R is an homomorphism of groups from $\widetilde{\mathrm{Homeo}}_T$ onto Homeo_T.

Proof: Let $g = G+\psi$. Write $g^{-1} = G^{-1} + \bar{\psi}$. We want to show that $\bar{\psi}(u) \to 0$ for $u \to +\infty$; $g^{-1}\circ g(u) = u = G^{-1}(G+\psi) + \bar{\psi}(G+\psi)$. So, $\bar{\psi}(G+\psi) = u - G^{-1}(G+\psi)$.

Now, the fact that $(G+\psi)(u) \to +\infty$ for $u \to +\infty$ and that G^{-1} is uniformly continuous, imply that:

$$G^{-1}(G+\psi)(u) = G^{-1}\circ G(u) + \tilde{\varphi}(u) = u + \tilde{\varphi}(u) \quad \text{with } \tilde{\varphi} \to 0 \text{ as } u \to +\infty.$$

Now, $\bar{\psi}(u) = \tilde{\psi}(g^{-1}(u))$, and because $g^{-1}(u) \to +\infty$ as $u \to +\infty$, we have $\bar{\psi}(u) \to 0$ for $u \to +\infty$.

Let now $g_1, g_2 \in \widetilde{\mathrm{Homeo}}_T$ with the decomposition $g_1 = G_1+\psi_1$ $g_2 = G_2+\psi_2$. Then $g_1\circ g_2(u) = G_1\circ(G_2(u)+\psi_2(u)) + \psi_1(g_2(u))$.
Now, as above, it is easy to see that

$$G_1\circ(G_2+\psi_2) = G_1\circ G_2+\tilde{\psi}_2 \quad \text{for a function } \tilde{\psi}_2,\ \tilde{\psi}_2(u) \to 0 \text{ as } u \to +\infty.$$

Also, we have $\psi_1\circ g_2(u) \to 0$ for $u \to +\infty$, and so

$$g_1\circ g_2(u) = G_1\circ G_2(u) + \psi(u), \quad \psi \to 0 \text{ for } u \to +\infty.$$

Remark. $\operatorname{Ker} R = \{g(u) = u+\psi(u) \mid \psi \to 0 \text{ for } u \to +\infty\}$ is a subgroup of $\widetilde{\mathrm{Homeo}}_T$, as well as Homeo_T. In fact, it follows from Lemma 3 that each $f \in \widetilde{\mathrm{Homeo}}_T$ may be uniquely written as:

$$f = R(f)\circ\alpha(f) = \beta(f)\circ R(f) \quad \text{with } \alpha(f),\beta(f) \in \operatorname{Ker} f.$$

So that $\widetilde{\mathrm{Homeo}}_T$ is generated by Homeo_T and $\operatorname{Ker} R$.

Corollary 4. The asymptotic translation number $\rho(f)$ in invariant by conjugacy in $\widetilde{\mathrm{Homeo}}_T$.

Proof: Let $H, f \in \widetilde{\mathrm{Homeo}}_T$. Then

$$R(H \circ f \circ H^{-1}) = R(H) \circ R(f) \circ R(H)^{-1} \quad \text{by Lemma 3.}$$

So, it follows that $\rho(H \circ f \circ H^{-1}) = \rho(f)$ by the invariance of the usual translation number under conjugacy in Homeo_T.

We call a <u>contraction</u> of $\mathbb{R}^+$ any homeomorphism of $\mathbb{R}^+$ such that $f(z) < z$ for all $z > 0$. Each contraction of $\mathbb{R}^+$ is conjugate to T, when restricted to $\mathbb{R}^+ - \{0\}$. So, the previous considerations may be translated to any contraction f. We will obtain relative to f the notions of space of homeomorphisms, number of translations, etc.

<u>Definitions</u>. Let f be a contraction of $\mathbb{R}^+$. We denote by Homeo_f the group of homeomorphisms of $\mathbb{R}^+$, preserving 0 and commuting with f. We denote by $\widetilde{\mathrm{Homeo}}_f = h \circ \widetilde{\mathrm{Homeo}}_T \circ h^{-1}$ the conjugate group of $\widetilde{\mathrm{Homeo}}_T$ by any conjugacy h between T and $f|(\mathbb{R}^+ - \{0\})$. (It follows from the Lemma 3 that $\widetilde{\mathrm{Homeo}}_f$ depends only on f and is independent of the choice of h). In the same way, we may define a <u>retraction</u> $R_f\colon \widetilde{\mathrm{Homeo}}_f \to \mathrm{Homeo}_f$ and we see that $\widetilde{\mathrm{Homeo}}_f$ is generated by Homeo_f and $\mathrm{Ker}\, R_f$. We define now the <u>relative asymptotic number of translation</u> $\rho(g;f)$ for any $g \in \widetilde{\mathrm{Homeo}}_f$ by the formula

$$\rho(g;F) = \rho(h^{-1} g\, h),$$

where h is any conjugacy of T with $f|(R^+ - \{0\})$. Again this number is well-defined.

All these notions are invariant by topological conjugacies. This assertion means that if h is a topological conjugacy between the two contractions f and g $(g = h f h^{-1})$, then $\mathrm{Homeo}_g = $ $= h\, \mathrm{Homeo}_f\, h^{-1}$, $\widetilde{\mathrm{Homeo}}_g = h\, \widetilde{\mathrm{Homeo}}_f\, h^{-1}$, $h \circ R_f = R_g \circ h$ and if $f' \in \widetilde{\mathrm{Homeo}}_f$ and $g' = h f' h^{-1}$, then $\rho(f';f) = \rho(g';g)$.

<u>Remark</u>. It is possible to define $\widetilde{\mathrm{Homeo}}_f$, R_f, $\rho(f;g)$ directly, that is, without making use of some conjugacy h. We present below a direct formula for $\rho(g;f)$. For $\mathrm{Ker}\, R_f$, for example, it is easy to verify that $g \in \mathrm{Ker}\, R_f$ if and only if $f^{-n} \circ g \circ f^n(z) - z \to 0$ uni-

formly if $z \in K$, K some compact of $\mathbb{R}^+ - \{0\}$, containing at least a fundamental domain of f (that is, an interval $[z_o, f(z_o)]$ for some $z_o \neq 0$).

<u>Lemma 5.</u> <u>Let</u> $g \in \widetilde{\text{Homeo}}_f$, <u>where</u> f <u>is some contraction of</u> $\mathbb{R}^+$. <u>Let</u> $z \in \mathbb{R}^+ - \{0\}$. <u>For every</u> $n \in \mathbb{N}$, <u>let</u> $m(n)$ <u>be the number such that</u> $f^{m(n)}(z) \leq g^n(z) < f^{m(n)+1}(z)$. <u>Then</u>

$$\rho(g;f) = \lim_{n\to\infty} \frac{m(n)}{n}.$$

<u>Proof</u>: Transport the couple (g,f) from $\mathbb{R}^+ - \{0\}$ to $\mathbb{R}$, to a couple $(\tilde{g}, T)$ by some conjugacy. We have by definition $\rho(g;f) = \rho(\tilde{g})$, and for the numbers n, $m(n)$ defined in the lemma

$$u + m(n) \leq \tilde{g}^n(u) < u + m(n) + 1.$$

So that $m(n) = [\tilde{g}^n(u) - u]$, where $[\alpha]$ means the entire part of the number α. Now

$$\rho(\tilde{g}) = \lim_n \frac{\tilde{g}^n(u)}{n} = \lim_n \frac{\tilde{g}^n(u) - u}{n} = \lim_n \frac{[\tilde{g}^n(u) - u]}{n} = \lim_n \frac{m(n)}{n}.$$

<u>Remark</u>. It is easy to generalize the previous lemma as follows. Let $K \subset \mathbb{R}^+ - \{0\}$ be any compact and $N \in \mathbb{N}$. Suppose that there exist sequences $\{z_i\}$, $\{\bar{z}_i\}$ in K and $\{m_i\}$, $\{n_i\}$ in $\mathbb{N}$, with $n_i \to +\infty$ such that, for every i

$$f^{m_i}(z_i) \leq g^{n_i}(\bar{z}_i) \leq f^{m_i+N}(z_i).$$

Then,

$$\rho(g;f) = \lim_{i\to+\infty} \frac{m_i}{n_i}.$$

Lemma 5 corresponds to the case $z_i \equiv \bar{z}_i \equiv z$ and $N = 1$. To have the condition on (z_i), $(\bar{z}_i)$ in some compact K satisfied, it suffices to have $(z_i)_i \to z \neq 0$ and $(\bar{z}_i)_i \to \bar{z} \neq 0$, and then take the restriction to $i \geq i_o$ for some i_o.

We give now an important example, used in the last part of this paper:

<u>Lemma 6. Let</u> f, g <u>be two homeomorphisms of</u> $\mathbb{R}^+$, <u>where</u> f <u>is a contraction of the form</u> $f(z) = \lambda z + o(z)$, <u>of class</u> C^2, <u>with</u> $0 < \lambda < 1$, <u>and</u> $g(z) = \mu z + o(z)$, <u>with</u> $\mu > 0$. <u>Then</u> $g \in \widetilde{\text{Homeo}}_f$ <u>and</u> $\rho(g;f) = \frac{\text{Log }\mu}{\text{Log }\lambda}$.

<u>Proof:</u> The map f is hyperbolic of class C^2, so there exists an homeomorphism $h_o(z) = z + o(z)$ such that

$$f_1(z) = h_o\, f\, h_o^{-1}(z) = \lambda z \qquad \text{(see [4] for example).}$$

Now $g_1(z) = h_o\, g\, h_o^{-1}(z) = \mu z + o(z)$ as q.

The map $z \to \lambda z$ is conjugate to T by $h_1(z) = \frac{\text{Log } z}{\text{Log }\lambda}$.

Now $\tilde{g} = h_1 \circ g_1 \circ h_1^{-1} = \frac{\text{Log } z}{\text{Log }\lambda}[\mu z(1+\varepsilon(z))] \circ h_1^{-1}$ where $\varepsilon(z) \to 0$ for $z \to 0$. It results from this

$$\tilde{g}(u) = u + \frac{\text{Log }\mu}{\text{Log }\lambda} + \psi(u)$$

where

$$\psi(u) = \frac{\text{Log}(1+\varepsilon \circ h_1^{-1}(u))}{\text{Log }\lambda} \to 0 \quad \text{for} \quad u \to +\infty.$$

So we have that $g \in \widetilde{\text{Homeo}}_f$ and that $\rho(g;f) = \frac{\text{Log }\mu}{\text{Log }\lambda}$.

<u>Remarks</u>

1) The hypothesis that f be of class C^2 can be some what weakened. It is enough to assume that Df is a Lipschitz map (see [4]).

2) The space $\widetilde{\text{Homeo}}_f$ depends on f in general, but it is the same for all C^2 hyperbolic contractions. This follows easily from Lemma 6: if $f(z) = f_\lambda(z) + o(z)$ with $f_\lambda(z) = \lambda \cdot z$, then $f_\lambda \in \widetilde{\text{Homeo}}_f$ and so $\widetilde{\text{Homeo}}_f = \widetilde{\text{Homeo}}_{f_\lambda}$. But we have trivially that $\widetilde{\text{Homeo}}_{f_\lambda} = \widetilde{\text{Homeo}}_{f_\mu}$ for any λ, μ, $0 < \lambda, \mu < 1$. Also, the groups $\text{Ker } R_f$ are the same for all hyperbolic contractions f. In fact, we have that $\text{Ker } R_f = \{g$, homeomorphism of $\mathbb{R}^+ \mid g(z) = z + o(z)\}$. Finally if $f_1(z) = \lambda_1 z + o(z)$ and $f_2(z) = \lambda_2 z + o(z)$ are two C^2 hyperbolic contractions (λ_1 and λ_2 with norm less than one) and if $g \in \widetilde{\text{Homeo}}_{f_1} = \widetilde{\text{Homeo}}_{f_2}$, we have $\rho(g;f_1) = \frac{\text{Log }\lambda_2}{\text{Log }\lambda_1} \cdot \rho(g;f_2)$.

II. Topological invariant for saddle-node families

As indicated in the introduction we consider a more general situation than just the differentiable saddle-node bifurcation. In this part, we let $\mathbb{R}^2_+ = \mathbb{R}^+ \times \mathbb{R}$, with coordinates $\mu \in \mathbb{R}^+$ and $x \in \mathbb{R}$. A μ-family of flows on the line will be denoted by $X_\mu(t)$ and also by $X_\mu(t,x)$ when we need to indicate the initial point $x \in \mathbb{R}$. We also let $\ell_- = \{(0,x) \subset \mathbb{R}^2_+ \mid x < 0\}$ and $\ell_+ = \{(0,x) \mid x > 0\}$.

Definition. A regular family (of topological flows) $X_\mu(t)$ will be a 1-parameter family of topological flows of the line, with the parameter $\mu \in \mathbb{R}^+$, such that

(a) There exists a unique stationary point (at $\mu = x = 0$) and for $(\mu,x) \neq (0,0)$, $X_\mu(t,x) > x$ if $t > 0$.

(b) There exist two topological segments ℓ, r, with extremities in ℓ_- and ℓ_+, respectively, transversal to the horizontal lines μ = constant, such that the time $t(\mu)$ necessary to go from ℓ to r, for the value of the parameter $\mu > 0$, is a strictly discreasing function of μ. More precisely, we want to say that ℓ, r are given as graphs $x = \ell(\mu)$, $x = r(\mu)$, with $\ell(0) < 0$ and $r(0) > 0$. The time $t(\mu)$ is given by $X_\mu(t(\mu),\ell(\mu)) = r(\mu)$.

Remarks: A saddle-node bifuraction for vector fields, restricted to $\mu \geq 0$, as introduced in [1], is of course a regular family of flows. The condition (b) is not a consequence of condition (a) in general. (F. Dumortier showed us an example of a family verifying (a) for which there is no pair of segments ℓ, r as in condition (b)).

Now, let $F_\mu = X_\mu(1)$, the time-one family of homeomorphisms, of a regular family of flows $X_\mu(t)$. We want to show that the flow $X_o(t)$ is a topological invariant of F_μ, that is, depends only on F_μ. Recall that a topological conjugacy between two families F_μ, $\bar{F}_\mu$ is a homeomorphism $H(\mu,x) = (\varphi(\mu),h_\mu(x))$ of $\mathbb{R}^2_+$ such that

$\bar{F}_{\varphi(\mu)} \circ h_\mu = h_\mu \circ F_\mu$.

Precisely, we have the following result:

Theorem 7. Let $X_\mu(t)$, $\bar{X}_\mu(t)$ be two regular families of flows. Then, if $H = (\varphi, h_\mu)$ is a topological conjugacy of $F_\mu = X_\mu(1)$ with $\bar{F}_\mu = \bar{X}_\mu(1)$, the homeomorphism h_o is a conjugacy of $X_o(t)$ with $\bar{X}_o(t)$ $(\bar{X}_o(t, h_o(x)) = h_o X_o(t,x))$. If a family F_μ of homeomorphisms imbeds in a regular family of flows, all the possible imbeddings coincide for $\mu = 0$ (i.e., if $X_\mu(t)$, $\bar{X}_\mu(t)$ are two regular families such that $F_\mu = X_\mu(1) = \bar{X}_\mu(1)$, then $X_o(t) = \bar{X}_o(t)$).

Remark. If one notes that any topological conjugate of a regular family is again a regular family, it is immediate that the two assertions of the theorem are equivalent.

This result is a simple generalization of [1]. Here, we want just to present a new proof of it, based on an interpretation of the flow $X_o(t)$ in terms of asymptotic numbers of translation. (See Theorem 9 below).

These numbers will be construct, using a pair of adapted foliations for F_μ (see the definition below). We begin with a simple example of such a pair of foliations strongly associated to the flow $X_\mu(t)$.

So, let $X_\mu(t)$ be a regular family of flows, and ℓ, r, a pair of transversal segments choosen as in the definition. We can define two topological foliations $\mathfrak{F}_L^o$ and $\mathfrak{F}_R^o$ just by "translating" ℓ and r using the flow $X_\mu(t)$. This means that if $X(t) = (\mu, X_\mu(t))$, we define the leaves of $\mathfrak{F}_L^o$ to be the lines $X(t,\ell)$ where $t \in \mathbb{R}$, and the leaves of $\mathfrak{F}_R^o$ to be the lines $X(t,r)$, $t \in \mathbb{R}$. The foliation $\mathfrak{F}_L^o$ is defined on $\mathbb{R}_+^2 - \ell_+$ and is transversal to the horizontal lines $\{\mu = \text{constant}\}$ for $\mu > 0$ and also to ℓ_-; the foliation $\mathfrak{F}_R^o$ is defined on $\mathbb{R}_+^2 - \ell_-$, transversal to the horizontal lines $\{\mu = \text{constant}\}$, $\mu > 0$, and to ℓ_+. Finally, $\mathfrak{F}_L^o$ and $\mathfrak{F}_R^o$ are mutually transversal on $\mathbb{R}^2_+ - (\ell_- \cup \ell_+)$ as a consequence of condition (b) of the definition. (Above, transversal means <u>topologically transversal</u>).

Now, we are going to interpret the flow $X_o(t)$, on ℓ_-, in terms of numbers of translation. The same interpretation can be made for $X_o(t)$ on ℓ_+. For each $x \in \ell_-$, let L^o_x be the $\mathfrak{F}^o_L$-leaf through x. This leaf is given by the graph of a function $L^o_x(\mu)$ such that $L^o_x(0) = x$. Take two points $x,y \in \ell_-$. We define a map $h_{xy}(\mu): \mathbb{R}^+ \circlearrowleft$ by the condition that $(\mu, L_x(\mu))$ and $(h_{xy}(\mu)),$ $L_y(h_{xy}(\mu)))$ belong to the same leaf of $\mathfrak{F}^o_R$. Now, the function $t(\mu)$ of condition (b) may be seen as an homeomorphism of $\mathbb{R}^+ - \{0\}$ to $\mathbb{R}$ (which can be extended by $t(0) = +\infty$). If we take the conjugacy of the map $h_{xy}(\mu)$ by the homeomorphism $u = t(\mu)$ we obtain:

$$\tilde{h}^o_{xy}(u) = t \circ h_{xy} \circ t^{-1}(u) = u + t(x,y)$$

where $t(x,y)$ is the time to go from x to y by the flow $X_o(t)$; i.e., $X_o(t(x,y),x) = y$. See fig. 1.

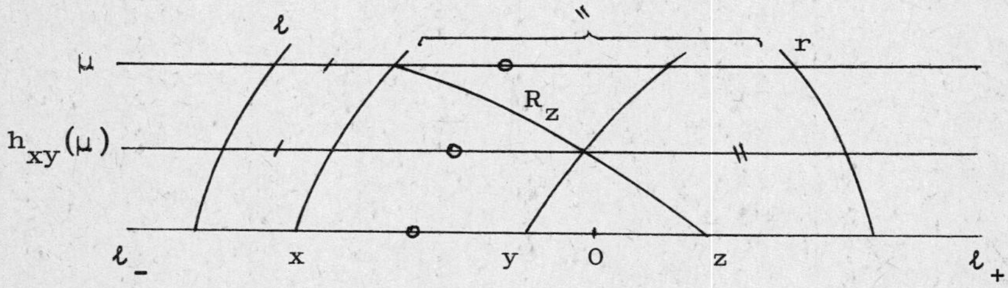

(o : time equal to $t(x,y)$; / , \\ : equal times.)

Figure 1

So, the time $t(x,y)$ is interpreted as the number of translation of $\tilde{h}_{xy}$. The idea, now, is to generalize somewhat the above construction to get rid of the dependence on $X_\mu(t)$.

<u>Definition</u>. Let $X_\mu(t)$ be a regular family and $F_\mu = X_\mu(1)$. <u>An</u>

adapted pair $(\mathfrak{F}_L,\mathfrak{F}_R)$ of foliations for F_μ is a pair of topological foliations, invariant by $F = (\mu,F_\mu)$ and such that:

(a) $\mathfrak{F}_R$ is defined on $\mathbb{R}^2_+ - \ell_-$, topologically transversal to the horizontal lines (for $\mu > 0$) and to ℓ_+,

(b) $\mathfrak{F}_L$ is defined on $\mathbb{R}^2_+ - \ell_+$, topologically transversal to the horizontal lines (for $\mu > 0$) and to ℓ_-,

(c) $\mathfrak{F}_L$ and $\mathfrak{F}_R$ are topologically transversal on $R^2_+ - (\ell_- \cup \ell_+)$.

The above example $(\mathfrak{F}^o_L,\mathfrak{F}^o_R)$ shows that adapted pairs do exist. Note that the class of all adapted pairs for F_μ is naturally associated to F_μ in the following sense: if $H = (\varphi(\mu),h_\mu)$ is a topological conjugacy between he families F_μ and $\bar{F}_\mu$, and if $(\mathfrak{F}_L,\mathfrak{F}_R)$ is any adapted pair for F_μ, then $(H(\mathfrak{F}_L),H(\mathfrak{F}_R))$ is an adapted pair for $\bar{F}_\mu$.

For any adapted pair $(\mathfrak{F}_L,\mathfrak{F}_R)$, we can defined as above a set of maps $h_{xy}(\mu)$ for $x,y \in \ell_-$ (the projection of the $\mathfrak{F}_L$ - leaf L_x through x on the $\mathfrak{F}_L$ - leaf L_y through y along leaves of $\mathfrak{F}_R$, all $\mathfrak{F}_L$ - leaves being parametrized by $\mu \in \mathbb{R}^+$). Note that for $y > x$, we have $h_{xy}(\mu) < \mu$ and so h_{xy} is a contraction of $\mathbb{R}^+$. For the special choice $y = F_o(x)$ we write $h_{xF_o(x)}(\mu) = T_x(\mu)$. All the facts we state here for ℓ_- may also be stated for ℓ_+. So we speak only about ℓ_- and leave the translation to ℓ_+ to the reader.

Our interpretation of $t(x,y)$ depends on the following proposition:

<u>Proposition 8</u>. <u>Let</u> $(\mathfrak{F}_L,\mathfrak{F}_R)$ <u>be a pair of adapted foliations for</u> F_μ <u>and any</u> $x_o \in \ell_-$. <u>Then, there exists a homeomorphism</u> $h\colon \mathbb{R}^+-\{0\} \to \mathbb{R}$ (<u>we set</u> $h(0) = +\infty$), <u>such that if</u> $\tilde{h}_{xy} = h \circ h_{xy} \circ h^{-1}$, <u>we have</u> $\tilde{T}_{x_o} = T$ <u>and, for any</u> $x,y \in \ell_-$, $\tilde{h}_{xy}(u) = u + t(x,y) + \psi_{xy}(u)$, <u>where</u> $\psi_{xy}(u)$ <u>is a continuous function in</u> (x,y,u), <u>such that, for</u> x, y <u>fixed</u>, $\psi_{xy}(u) \to 0$ <u>for</u> $u \to +\infty$. (<u>Recall that</u> $t(x,y)$ <u>is the time to go</u>

<u>from</u> x <u>to</u> y <u>by the flow</u> $X_o(t)$).

Before giving the proof of Proposition 8, we want to deduce from it our interpretation of Theorem 7 in terms of asymptotic numbers of translation.

<u>Theorem 9</u>. <u>Let</u> $F_\mu = X_\mu(1)$ <u>be a family of flows</u>. <u>Let</u> $(\mathfrak{F}_L,\mathfrak{F}_R)$ <u>be any pair of adapted foliations for</u> F_μ <u>and consider a point</u> $x_o \in \ell_-$. <u>Then the relative asymptotic number of translation</u> $\rho(h_{xy};T_{x_o})$ <u>is independent of the choice of</u> $(\mathfrak{F}_L,\mathfrak{F}_R)$ <u>and</u> x_o. <u>We denote it by</u> $\rho_{F_\mu}(x,y)$. (<u>By the above facts for</u> $(\mathfrak{F}_R^o,\mathfrak{F}_L^o)$ <u>in Proposition 8, it follows that</u> $\rho_{F_\mu}(x,y) = t(x,y)$ <u>which is the desired interpretation</u>). <u>If</u> $H = (\varphi(\mu),h_\mu)$ <u>is a conjugacy between two families</u> F_μ, $\bar{F}_\mu$ <u>of time-one maps, corresponding to regular families of flows</u> $X_\mu(t)$, $\bar{X}_\mu(t)$, <u>we have</u>:

$$\rho_{F_\mu}(x,y) = \rho_{\bar{F}_\mu}(h_o(x),h_o(y)).$$

(<u>So</u>, h_o <u>conjugates</u> $X_o(t)$ <u>and</u> $\bar{X}_o(t)$ <u>on</u> ℓ_-).
<u>The same results are also true on</u> ℓ_+.

<u>Proof of the Theorem 9</u>: Let x_o, $(\mathfrak{F}_L,\mathfrak{F}_R)$ as in the theorem. From Proposition 8 we know that, for some conjugacy h between T_{x_o} and T, we have

$$\tilde{h}_{xy}(\mu) = u + t(x,y) + \psi_{xy}(u).$$

It follows from our study in part I that:

$$\rho(h_{xy},T_{x_o}) = t(x,y).$$

This formula shows that $\rho(h_{xy},T_{x_o})$ is independent of the choice of $(\mathfrak{F}_L,\mathfrak{F}_R)$ and x_o, so it depends only on F_μ.
If now $H = (\varphi(\mu),h_\mu)$ is a conjugacy from F_μ to $\bar{F}_\mu$, $(H\mathfrak{F}_L,H\mathfrak{F}_R)$ is a pair of adapted foliations for $\bar{F}_\mu$. If $x_o \in \ell_-$ and h is a conjugacy between T_{x_o} and T, then $h\circ\varphi^{-1}$ is a conjugacy between $\bar{T}_{h_o(x_o)}$ and T and for it, we have precisely that $\tilde{h}_{h_ox,h_oy}(\mu) =$

$= \tilde{h}_{xy}(\mu)$. (Here $\bar{h}_{xy}$ is the map associated to $\bar{F}_\mu$ and the pair of foliations $(H\mathcal{F}_L, H\mathcal{F}_R)$, and $\tilde{\bar{h}}_{h_o x, h_o y}$ is its conjugate by $h \cdot \varphi^{-1}$). So we obtain the result: $\rho_{F_\mu}(x,y) = \rho_{\bar{F}_\mu}(h_o x, h_o y)$.

<u>Proof of Proposition 8</u>: Let $(\mathcal{F}_R, \mathcal{F}_L)$ be any adapted pair for F_μ. Let also $(\mathcal{F}_R^o, \mathcal{F}_L^o)$ be the special pair introduced in the beginning, which is invariant by the flow X_μ. We want to find a choice of homeomorphism h such that $\tilde{h}_{xy}$, associated to $(\mathcal{F}_R, \mathcal{F}_L)$ and h, differ from the translation $u + t(x,y)$ just by some function $\psi_{xy}(u)$ with $\psi_{xy} \to 0$ as $u \to +\infty$. First, in Lemma 10, we see that we can suppose that $\mathcal{F}_R^o = \mathcal{F}_R$:

<u>Lemma 10</u>. <u>Let</u> $\mathcal{F}_R$, $\mathcal{F}_R^o$ <u>be two invariants foliations by</u> F_μ, <u>defined on</u> $\mathbb{R}_+^2 - \ell_-$ (<u>and transversal to</u> ℓ_+ <u>and to the horizontal lines</u>, <u>for</u> $\mu > 0$). <u>Then, there exists a continuous family of homeomorphisms of the line</u> h_μ, $\mu \in \mathbb{R}^+$, <u>commuting with</u> F_μ ($F_\mu \circ h_\mu = h_\mu \circ F_\mu$ <u>for all</u> μ), <u>such that</u> $h_o = \mathrm{Id}$ <u>and such that</u> $H = (\mu, h_\mu)$ <u>sends</u> $\mathcal{F}_R^o$ <u>to</u> $\mathcal{F}_R$.

<u>Proof of Lemma 10</u>: We denote by R_z, R_z^o the leaves of $\mathcal{F}_R$, $\mathcal{F}_R^o$ through the point $z \in \ell_+$. These leaves are graphs of functions $R_z(\mu)$ and $R_z^o(\mu)$. Now, the conditions of the lemma impose that $h_\mu(R_z^o(\mu)) = R_z(\mu)$ for every $z \in \ell_+$ and $\mu \in \mathbb{R}^+$. This formula defines a homeomorphism $H = (\mu, h_\mu)$ in $\mathbb{R}_+^2 - \ell_-$ which commutes with F_μ. (This point follows from the F_μ-invariance of $\mathcal{F}_R^o$ and $\mathcal{F}_R$). It remains to show that H is a continuous extension of $h_o = \mathrm{Id}$ on ℓ_-. At this point we need to know that $F_\mu = X_\mu(1)$ for a topological flow X_μ. For a point $(\mu,x) \in \mathbb{R}_+^2 - \ell_-$ we have

$$(\mu,x) = (\mu, R_z^o(\mu)) \quad \text{for some } z \text{ in } \ell_+ .$$

We define a continuous function $t(\mu,x)$ in $\mathbb{R}_+^2 - \ell_-$ by: $t(\mu,x)$ is the time to go from $R_z^o(\mu)$ to $R_z(\mu)$ by X_μ. This function verifies $t(0,x) = 0$ for $x \in \ell_+$ and $t(\mu, F_\mu(x)) = t(\mu,x)$. Clearly

$$h_\mu(x) = X_\mu(t(\mu,x),x). \tag{1}$$

We will use this formula to show the continuity of H on ℓ_-. Let W be a fundamental domain for $\mathfrak{F}_R^o$, that is, a band between two leaves $R^o_{z_o}$, $R^o_{F_o(z_o)}$ (see fig. 2). For each $(\mu,x) \in \mathbb{R}^2_+ - \ell_-$, there exists an integer $n(\mu,x)$ such that $F_\mu^{n(\mu,x)}(x) \in W$. Consider a point $(0,x)$ in ℓ_- and a sequence $(\mu_i,x_i) \to (0,x)$. We can write:

$$h_{\mu_i}(x_i) = X_{\mu_i}(t(\mu_i,x_i),x_i) = X_{\mu_i}(t(\mu_i,F_{\mu_i}^{n(\mu_i,x_i)}(x_i)),x_i).$$

Now, the sequence $(\mu_i,F_{\mu_i}^{n(\mu_i,x_i)}(x_i)) \in W$ and the continuity of $t(\mu,x)$ and the fact that $\mu_i \to 0$ imply that:

$$t(\mu_i,F_{\mu_i}^{n(\mu_i,x_i)}(x_i)) \to 0 \quad \text{as} \quad i \to +\infty.$$

By continuity of $X_\mu(t)$, it follows that $h_{\mu_i}(x_i) \to X_o(0,x) = x$. This proves the continuity of $H = (\mu,h_\mu)$ on ℓ_-.

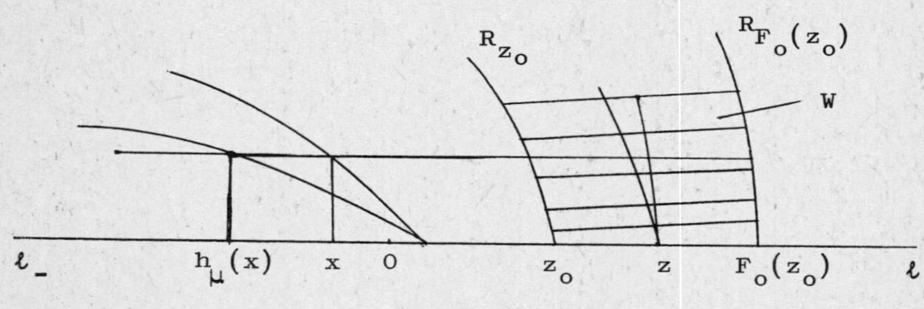

Figure 2

If H is the homeomorphism found in Lemma 10, we can replace the flow X_μ by its conjugate by H and $(\mathfrak{F}_L^o,\mathfrak{F}_R^o)$ by $(H\mathfrak{F}_L^o,H\mathfrak{F}_R^o)$. The conjugate flow is again a regular family in which F_μ imbeds (because h_μ commutes with F_μ). The advantage now is that $H\mathfrak{F}_R^o = \mathfrak{F}_R$.

So we simple write X_μ, $(\mathfrak{F}_L^o,\mathfrak{F}_R^o)$ for these conjugates, with the assumption that $\mathfrak{F}_R^o = \mathfrak{F}_R$. We recall that the conjugacy $t(\mu)$

(time to go from a given leaf ℓ of $\mathfrak{F}_L^o$ to a given leaf r of $\mathfrak{F}_R^o$) brings the maps h_{xy}^o associated to $(\mathfrak{F}_L^o,\mathfrak{F}_R^o)$ to the translations $u + t(x,y)$. To show Proposition 8, we have first to show the following lemma.

<u>Lemma 11.</u> <u>For each</u> $x \in \ell_-$, <u>let</u> s_x <u>be the map defined by the condition that</u> $(\mu, L_x^o(\mu))$ <u>and</u> $(s_x(\mu), L_x(s_x(\mu)))$ <u>belong to the same leaf of</u> $\mathfrak{F}_R = \mathfrak{F}_R^o$. <u>Then:</u>

$t \circ s_x \circ t^{-1}(u) = \tilde{s}_x(u) = u + \psi_{xy}(u)$, <u>where</u> $\psi_{xy}(u)$ <u>is continuous in</u> (x,y,u) <u>and</u> $\psi_{xy}(u) \to 0$ <u>with</u> $u \to +\infty$.

<u>Proof of the Lemma 11:</u> To study $\tilde{s}_x(u)$, we take the homeomorphism of $\mathbb{R}_+^2 - (\ell_- \cup \ell_+)$ to $\mathbb{R}^2$ given by $H(\mu,x) = (t(\mu),x)$. Denote by $\tilde{\mathfrak{F}}_L, \tilde{\mathfrak{F}}_L^o, \tilde{\mathfrak{F}}_R^o$ the conjugate foliations and by $\tilde{L}_x, \tilde{L}_x^o$ the image of the leaves L_x, L_x^o. The leaves $\tilde{L}_x$, $\tilde{L}_x^o$ are graphs of functions $\tilde{L}_x(u)$, $\tilde{L}_x^o(u)$, $u \in \mathbb{R}$. Because L_x and L_x^o pass through the point $(0,x)$, we have:

$$\tilde{L}_x(u) - \tilde{L}_x^o(u) \to 0 \quad \text{for} \quad u \to +\infty.$$

So, given $\delta > 0$, there exists $u_o(\delta)$ such that for

$$u \geq u_o(\delta), \quad \tilde{L}_x(u) \in [\tilde{L}_{x-\delta}^o(u), \tilde{L}_{x+\delta}^o(u)] \tag{1}$$

Now let $\tilde{R}(u)$ be the leaf of $\tilde{\mathfrak{F}}_R^o$ passing through the point $p(u) = (u, \tilde{L}_x^o(u))$. $\tilde{R}(u)$ is the graph of the function

$$y \to u + t(x,y).$$

Let $A(u) = \tilde{L}_{x-\delta}^o \cap \tilde{R}(u)$ and $C(u) = \tilde{L}_{x+\delta}^o \cap \tilde{R}(u)$. We complete the picture in a rectangle $\{A(u),B(u),C(u),D(u)\}$ defined by the conditions that the sides $[A,B]$ and $[C,D]$ are horizontal and $[A,D] \subset \tilde{L}_{x-\delta}^o$, $[B,C] \subset \tilde{L}_{x+\delta}^o$. See figure 3.

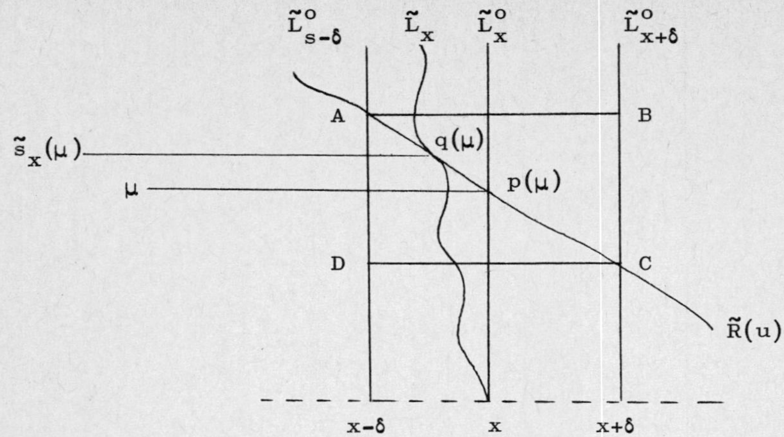

Figure 3

Now, because the function $y \to u + t(x,y)$ is decreasing with y (the foliation $\tilde{\mathcal{F}}^o_R$ is transversal to the horizontal lines), the part of $\tilde{R}(u)$ for $y \in [x-\delta, x+\delta]$ is entirely contained in the rectangle $\{A,B,C,D\}$. If $u \geq u_o(\delta) + t(x-\delta,x)$, the leaf $\tilde{L}_x$ must cut the sides $[AB]$ and $[CD]$ and do not cut the sides $[AD]$ and $[B,C]$. So, the unique intersection $q(u)$ between $\tilde{L}_x$ and $\tilde{R}(u)$ must belong to the rectangle $\{A,B,C,D\}$. Then, the vertical distance between $p(u)$ and $q(u)$, that is $|\tilde{s}_x(u)-u|$, satisfies

$$|\tilde{s}_x(u)-u| \leq -t(x,x-\delta) + t(x,x+\delta) = t(x-\delta,x+\delta)$$

if $u \geq u_o(\delta) + t(x-\delta,x)$.

Because $t(x,y)$ is continuous in (x,y), it follows that $\tilde{s}_x(u)-u \to 0$ for $u \to +\infty$. (Given $\varepsilon > 0$, you choose $\delta > 0$ such that $t(x-\delta,x+\delta) \leq \varepsilon$. If $u \geq u_o(\delta) + t(x-\delta,x)$, then $|\tilde{s}_x(u)-u| \leq \varepsilon$).

<u>End of the proof of Proposition 8</u>.

Let h_{xy} be the maps associated with the pair $(\mathcal{F}_L,\mathcal{F}_R)$, and h^o_{xy} with the pair $(\mathcal{F}^o_L,\mathcal{F}^o_R = \mathcal{F}_R)$. Then, using the notation of Lemma 11, we have:

$$h_{xy} = s_y \circ h^o_{xy} \circ s_x^{-1} . \qquad (1)$$

Now for the conjugacy by t we have:

$$\tilde{h}^o_{xy}(u) = u + t(x,y) \quad \text{and} \quad \tilde{s}_y(u), \tilde{s}_x^{-1}(u)$$

are of the form $u + \psi_{xy}(u)$ (Lemma 11). So, it follows from Lemma 3 that:

$$\tilde{h}_{xy} = u + t(x,y) + \psi_{xy}(u) \quad \text{with a} \quad \psi_{ny} \to 0 \quad \text{with} \quad \mu \to +\infty .$$

Now, let $x_o \in \ell_-$, $T_{x_o} = h_{x,F_o(x)}$. The map t conjugates $T^o_{x_o}$ with T but it does not conjugate T_{x_o} with T in general. So we have to modify this conjugacy t.

First, applying (1) to x_o and $F_o(x_o)$, we have:

$$T_{x_o} = s_{F_o(x_o)} \circ T^o_{x_o} \circ s^{-1}_{x_o} .$$

As a consequence of the F_μ-invariance of the foliations we have $s_{F_o(x_o)} = s_{x_o}$. So

$$T_{x_o} = s_{x_o} \circ T^o_{x_o} \circ s^{-1}_{x_o} . \qquad (2)$$

Now conjugate the expression in (2) with the homeomorphism t. We know that $\tilde{T}^o_{x_o} = T$, so $\tilde{T}_{x_o} = \tilde{s}_{x_o} \circ T \circ \tilde{s}^{-1}_{x_o}$. This means that $\tilde{s}^{-1}_{x_o}$ is a conjugacy between $\tilde{T}_{x_o}$ and T. So $h = \tilde{s}^{-1}_{x_o} \circ t$ will be a conjugacy from T_{x_o} to T. By this conjugacy, h_{xy} is send to $\tilde{\tilde{h}}_{xy} = \tilde{s}^{-1}_{x_o} \circ \tilde{h}_{xy} \circ \tilde{s}_{x_o}$ and so, by Lemma 3 again, we have:

$$\tilde{\tilde{h}}_{xy}(u) = u + t(x,y) + \psi_{xy}(u),$$

for some function ψ_{xy} tending to 0 with $u \to +\infty$. The conjugacy h is the desired one.

III. Topological invariant for saddle-connections in dimension 2.

We begin with a degenerate situation, where the invariant $\frac{\operatorname{Log}|\mu|}{\operatorname{Log}|\lambda|}$ is somewhat easier to interpret. So, let F be a diffeomorphism of class C^2 of some 2-dimensional manifold M^2, with two saddle fixed points p, q having a common separatrix $\gamma \subset W^u(p) \cap \cap W^s(q)$. We may also suppose that M^2 is oriented and that F preserves its orientation. Choose finally two separatrices $S_p \subset W^s(p)$ and $U_q \subset W^u(q)$, such that S_p and U_q are on the same side of γ. Let λ, μ, $0 < \lambda < 1 < \mu$ be the stable and unstable eigenvalues in p and q, respectively. We want to interpret the topological invariance of the ratio $\frac{\operatorname{Log} \mu}{\operatorname{Log} \lambda}$. Precisely, we have the following (see [1]):

<u>Theorem 12</u>. <u>Suppose that</u> F, F' <u>are two</u> C^2 <u>diffeomorphisms on manifolds</u> M^2, M'^2 <u>as above, with a pair of saddle fixed points having a common separatrix</u> (γ <u>a separatix for</u> F <u>from</u> p <u>to</u> q, γ' <u>a separatix for</u> F' <u>from</u> p' <u>to</u> q'). <u>Let</u> λ, μ <u>be the stable and unstable eigenvalues at</u> p <u>and</u> q, <u>respectively</u>, λ', μ' <u>the stable and unstable eigenvalues at</u> p' <u>and</u> q', <u>respectively</u>. <u>Suppose now that there exist neighborhoods</u> U <u>of</u> $\bar{\gamma}$ <u>and</u> U' <u>of</u> $\bar{\gamma}'$ <u>and a topological conjugacy</u> G <u>from</u> F_U <u>to</u> $F'_{U'}$. <u>Then</u> $\frac{\operatorname{Log} \mu}{\operatorname{Log} \lambda} = \frac{\operatorname{Log} \mu'}{\operatorname{Log} \lambda'}$.

Here, we want to interpret the ratio $\frac{\operatorname{Log} \mu}{\operatorname{Log} \lambda}$ <u>as a relative asymptotic number of translation</u>. The idea is to associate to F a pair of homeomorphisms of $\mathbb{R}^+$, $F|_{S_p}$ and $F|_{U_q}$, of hyperbolic type like in Lemma 6. But here we have $S_p \neq U_q$. So, to obtain a pair of homeomorphisms of the same line $S_p \simeq \mathbb{R}^+$, we will transport the map $F|_{U_q}$ from U_q to S_p using the following extra geometrical objet.

<u>Definition</u>. An <u>adapted system to the connection</u> γ <u>of</u> F will be a triple $S = (\mathfrak{F}_L, \mathfrak{F}_R, \sigma)$ where:

(a) $\mathfrak{F}_L$ is a topological F-invariant foliation, transversal to S_p and containing a neighborhood of γ in $W^u(p)$ as leaf,

(b) $\mathfrak{F}_R$ is a topological F-invariant foliation, transversal to

U_q and containing a neighborhood of γ in $W^s(q)$ as leaf,

(c) σ is a topological segment, homeomorphic to $\mathbb{R}^+$, with end point r in int γ and topologically transversal to $\mathfrak{F}_R$ and $\mathfrak{F}_L$.

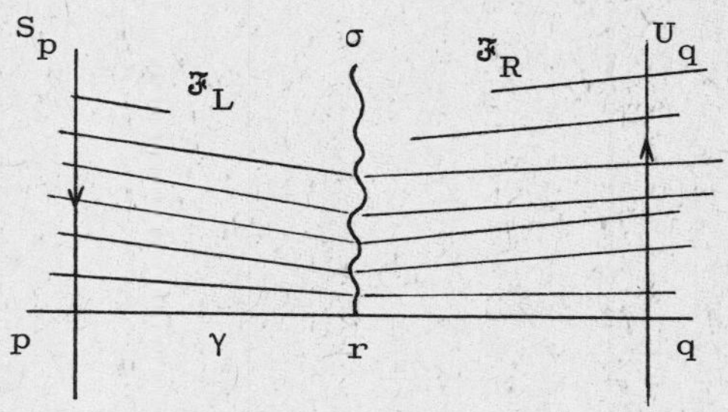

Figure 4

An easy consequence of the linearization theorem of Grobman-Hartman as formulated in [2],[3], is that there exists a neighborhood W_q of $[r,q] \subset \bar{\gamma}$ with coordinates (x,z) (z along U_q) such that the foliation $\mathfrak{F}_R$ defines a projection $\pi_R: W_q \to U_q$ with the property:

$$z(m) = \tau_k(m) + o(\pi_k(m)) \quad \text{for} \quad m \in W_q.$$

The same can be said for a neighborhood of p.

Now, for any adapted system S, we associate a local homeomorphism H_S from some neighborhood of q in U_q to some neighborhood of p in S_p, by following the leaf of $\mathfrak{F}_R$ from any point $m \in U_q$ (near enough q) to σ, and then following the leaf of $\mathfrak{F}_L$ through this point until we reach S_p.

It follows at once from the above remark about the linearization that we have:

<u>Lemma 13</u>. <u>For any adapted system</u> S, <u>the conjugate</u> g_S <u>of</u> $F|_{U_q}$ <u>by</u> H_S; $g_S = H_S \circ F|_{U_q} \circ H_S^{-1}$, <u>is a local</u> C^0 <u>homeomorphism of</u> S_p,

<u>defined in a neighborhood of</u> p <u>in</u> S_p, <u>of the form</u> $g_S(z) = \mu z + o(z)$.

Suppose that $g_S(z)$ is extended to all of S_p. Let $f = F|_{S_p}$. Then the pair (f, g_S) verifies the conditions of Lemma 6, and it follows that $\rho(g_S;f) = \frac{\text{Log }\mu}{\text{Log }\lambda}$. This formula shows that $\rho(g_S;f)$ doesn't depend on S, but just on F. Now the topological invariance of $\frac{\text{Log }\mu}{\text{Log }\lambda}$ follows from the naturality of the class of adapted systems. More precisely, suppose that G is a conjugacy like in Theorem 12 between F and F' in a neighborhood of the connection γ of F. Let S, be any adapted system for F $(S = (\mathcal{F}_L, \mathcal{F}_R, \sigma))$. Then $S' = G(S) = (G(\mathcal{F}_L), G(\mathcal{F}_R), G(\sigma))$ an adapted system for F' and $G|_{S_p}: S_p \to S_{p'}$ is a topological conjugacy of the pair (f, g_S) with the corresponding pair $(f', g_{S'})$ on $S_{p'}$ (here $g_{S'}$ is obtained from $H_{S'}$). The equality $\frac{\text{Log }\mu}{\text{Log }\lambda} = \frac{\text{Log }\mu'}{\text{Log }\lambda'}$ follows from the topological invariance of the relative asymptotic number of translation.

<u>Remark</u>. In [1], the topological invariance of $\frac{\text{Log }\mu}{\text{Log }\lambda}$ is obtained as follows. Let $\{r_i\}$, $r_i \notin \gamma$, be a sequence of points which converges to some $r \in \gamma$. Then, there exist sequences $\{m_i\}$, $\{n_i\}$ (after passing to subsequence of (r_i)) such that $f^{-m_i}(r_i)$ and $f^{n_i}(r_i)$ have limits on $S_p - \{p\}$ and $U_q - \{q\}$, respectively. Then it is shown in [1], that:

$$\frac{\text{Log }\mu}{\text{Log }\lambda} = - \lim_{i \to \infty} \frac{m_i}{n_i} .$$

If we take now an adapted system $S = (\mathcal{F}_L, \mathcal{F}_R, \sigma)$ such that $r_i \in \sigma$ (which is always possible by taking a subsequence of $\{r_i\}$) and if R_i is the projection of r_i on S_p along the leaves of $\mathcal{F}_L$, then $R_i \to p$ and $\bar{z}_i = g_S^{n_i}(R_i)$, $z_i = f^{-m_i}(R_i)$ have limits on $S_p - \{p\}$. So $\rho(g_S;f) = - \lim_{i \to \infty} \frac{m_i}{n_i}$ by the remark following Lemma 5.

<u>Our point here is just to interpret the formula</u> $\frac{\text{Log }\mu}{\text{Log }\lambda} = - \lim_i \frac{m_i}{n_i}$ <u>as a relative assymptotic number of translation.</u>

Consider now the generic situation, as in [1], of a quasi-transversal contact at r between $W^u(p)$ and $W^s(q)$. See figure 5 and [1] for more details.

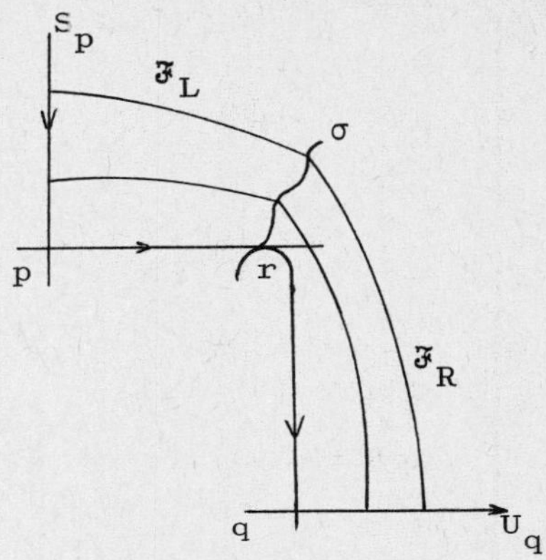

Figure 5

Again, in this generic situation we want to define an adapted system $(\mathfrak{F}_L, \mathfrak{F}_R, \sigma)$ with $\mathfrak{F}_L$ containing a neighborhood of $[p,r]$ in $W^s(p)$, $\mathfrak{F}_R$ containing a neighborhood of $[r,q]$ in $W^u(q)$ and σ passing through r. But now it is not true in general that g_S is of the form $g_S(z) = \mu z + o(z)$ and even that $g_S \in \widetilde{\text{Homeo}}_f$. So we have to add this last condition in the following definition.

<u>Definition</u>. An adapted system of F is a triple $S = (\mathfrak{F}_L, \mathfrak{F}_R, \sigma)$ defined as above, with the condition that $g_S \in \widetilde{\text{Homeo}}_f$.
(Recall that g_S is the conjugate of $F|_{U_q}$ by the transport along the leaves of $\mathfrak{F}_R$ from U_q to σ, and next along the leaves of $\mathfrak{F}_L$ from σ to S_p).

Now, for any adapted system the remark made after Lemma 13 is still valid: one can calculate $\rho(g_S;f)$ by taking a sequence $r_i \to r$, with $r_i \in \sigma$, using the fomrula

$$\rho(g_S;f) = -\lim_i \frac{m_i}{n_i}.$$

In [1], it is shown that we have the inequality

$$\frac{\text{Log } \mu}{\text{Log } \lambda} \leq - \lim_i \frac{m_i}{n_i} .$$

(The equality is realized when you take for example a differentiable σ, differentiably transversal to $W^u(p)$ and $W^s(q)$ in r). In the present context, we may translate this result as follows:

Lemma 14. Let F be a diffeomorphism exhibiting a quasi-transversal saddle connection as above. Then

$$\frac{\text{Log } \mu}{\text{Log } \lambda} = \text{Inf}\{\rho(g_S;f) \mid S, \text{ adapted system for } F\}.$$

The second number of the equality in Lemma 14 depends only on the conjugacy class of F, so the ratio $\frac{\text{Log } \mu}{\text{Log } \lambda}$ is a conjugacy invariant. The details are as in the degenerate situation we described above.

References

[1] S. Newhouse, J. Palis and F. Takens, Bifurcations and stability of families of Diffeomorphisms, Publ.Math.l'IHES, nº 57, p. 5-72.

[2] J. Palis, On the local structure of hyperbolic fixed points in Banach spaces, Anais Acad. Bras. Ciências 40 (1968), p.263-266.

[3] Ch. Pugh, On a theorem of P. Hartman, Amer. J. Math. 91 (1960), p.363.

[4] S. Sternberg, On local C^n contractions of the real line, Duke Math. J. 24 (1957), p. 97-102.

J. Palis

Instituto de Matemática Pura e Aplicada (IMPA)
Estrada Dona Castorina 110, Jardim Botânico
CEP 22460 - Rio de Janeiro, RJ - Brasil

R. Roussarie

Laboratoire de Topologie, ERA 0945 du CNRS
Departament de Mathématiques, UER MIPC
Université de Dijon
B.P. 138
21004. Dijon - France

ABELIAN INTEGRALS AND GLOBAL HOPF BIFURCATIONS

J.A. Sanders
Department of Mathematics
and Computer Science
Vrije Universiteit
Amsterdam, the Netherlands

R. Cushman
Mathematics Institute
Rijksuniversiteit Utrecht
Utrecht, the Netherlands

ABSTRACT : We give a detailed and simplified proof of a theorem of Yu. S. Il'yashenko which concerns the uniqueness of certain limit cycles. A slightly extended version of this theorem is then applied to a global Hopf bifurcation problem treated by Keener.

INTRODUCTION

In the first part of this paper we will explain and prove a result of Il'yashenko [1]. In the second part we apply this theorem to a global bifurcation problem treated by Keener [2]. Not only does Il'yashenko's theorem remove an unnecessary monotonicity assumption in Keener, but it does this in such a way that hardly any calculations are needed.

The proof that we give is independent of Il'yashenko's [3]. He uses local computations together with global information from Picard-Lefschetz theory [4] to obtain his results. On the other hand, we derive the Picard-Fuchs equation for the given family of elliptic curves (1.1) and then make a study of the global properties of the associated Riccati equation (1.6). Our method involves more calculations, but these computations give internal checks, which allow us to avoid trivial errors which mar Il'yashenko's papers.

In [1] Il'yashenko also treats cases with $\mathbb{Z}/2\mathbb{Z}$ and $\mathbb{Z}/3\mathbb{Z}$ symmetry. These cases have been worked out using our method, but will not be treated here.

We prove the following

THEOREM (Il'yashenko) :*Consider the family of real affine cubic curves*

$$\Gamma : y^2 = x^3 - 3x + s \qquad (1)$$

and let γ_s *be the compact component of* Γ *for* $s \in (-2,2)$ *(see Figure1). On* $[-2,2]$ *define the function*

$$\xi(s) = \int_{\gamma_s} xydx \Big/ \int_{\gamma_s} ydx. \qquad (2)$$

Then ξ *is strictly monotonic on* $[-2,2]$ *with range* $[-1,-5/7]$.

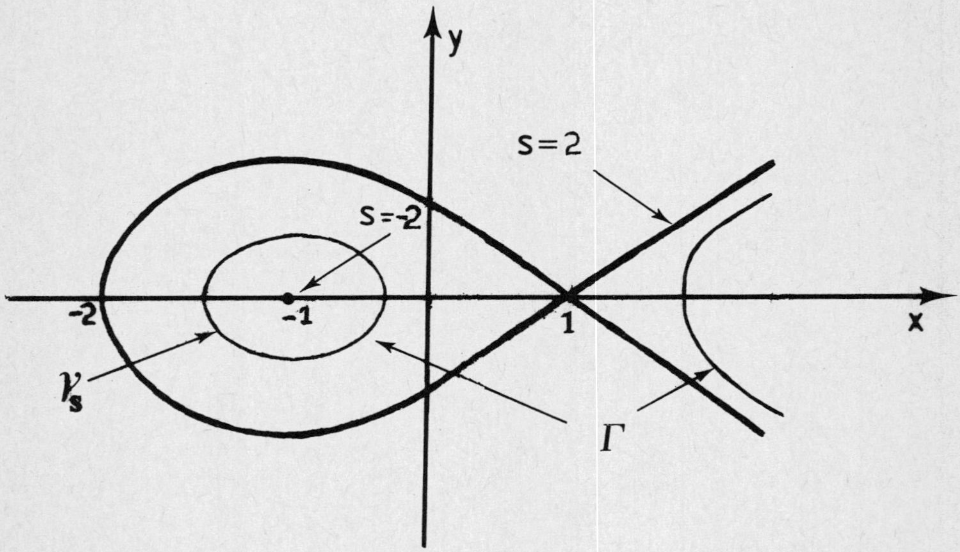

Figure 1. The family Γ of real affine cubic curves.

In [1] the proof of the above theorem contains two mistakes : 1) the expression $f(\gamma)$ in the first line of table 1 should read $(7\gamma^2-5)/\gamma$ and 2) the assertion in proposition 8 that the tangent to $M(=\xi)$ at $c = \pm 2$ ($c=s$) is parallel to the c axis is wrong. In fact $M(-2^+) = 1/24$ and $M'(2^-) = \infty$.

Postponing all calculations to §2 and §3, we give our proof that

ξ is strictly monotonic increasing. Consider the differentials

$$\alpha = ydx \text{ and } \beta = xydx$$

and their periods

$$\mathcal{A} = \int_{\gamma_s} \alpha \text{ and } \mathcal{B} = \int_{\gamma_s} \beta$$

over γ_s. Let

$$a = \frac{dx}{y} \text{ and } b = \frac{xdx}{y}.$$

Then a and b are differentials of the first and second kind [5] on the compact Riemann surface defined by complexifying and homogenizing Γ. Let

$$A = \int_{\gamma_s} a \text{ and } B = \int_{\gamma_s} b$$

be the period of a and b over γ_s. We may express $\mathcal{A}$ and $\mathcal{B}$ in terms of A and B as follows:

$$\begin{aligned} \mathcal{A} &= \frac{3s}{5} A - \frac{6}{5} B \\ \mathcal{B} &= -\frac{6}{7} A + \frac{3s}{7} B. \end{aligned} \qquad (3)$$

A calculation shows that

$$\xi(-2) = -1 \text{ and } \xi(2) = -5/7. \qquad (4)$$

By Stokes' theorem

$$\mathcal{A} = \int_{\gamma_s} ydx = \int_{D_s} dxdy$$

is the area of the region D_s enclosed by γ_s. Therefore $\mathcal{A} > 0$ for all $s \in (-2,2)$. Consequently, $\xi = {}^{\mathcal{B}}/_{\mathcal{A}}$ is continuous on $(-2,2)$. From (4) it follows that ξ is continuous on $[-2,2]$.

Now $\mathcal{A}$ and $\mathcal{B}$ satisfy the following system of differential equations

$$\begin{aligned} 6(s^2-4)\frac{d\mathcal{A}}{ds} &= 5s\mathcal{A} + 14\mathcal{B} \\ 6(s^2-4)\frac{d\mathcal{B}}{ds} &= 10\mathcal{A} + 7s\mathcal{B} \end{aligned} \qquad (5)$$

which is called the Picard-Fuchs equation associated to Γ. From (5) we obtain a Riccati equation for ξ

$$3(s^2-4)\frac{d\xi}{ds} = 5 + s\xi - 7\xi^2. \qquad (6)$$

We want to show that all solutions of (6) with boundary values given in (4) are strictly monotonic increasing. From (6) and the fact that the branches of $5 + s\xi - 7\xi^2 = 0$ are strictly monotonic increasing functions of s on (-2,2),

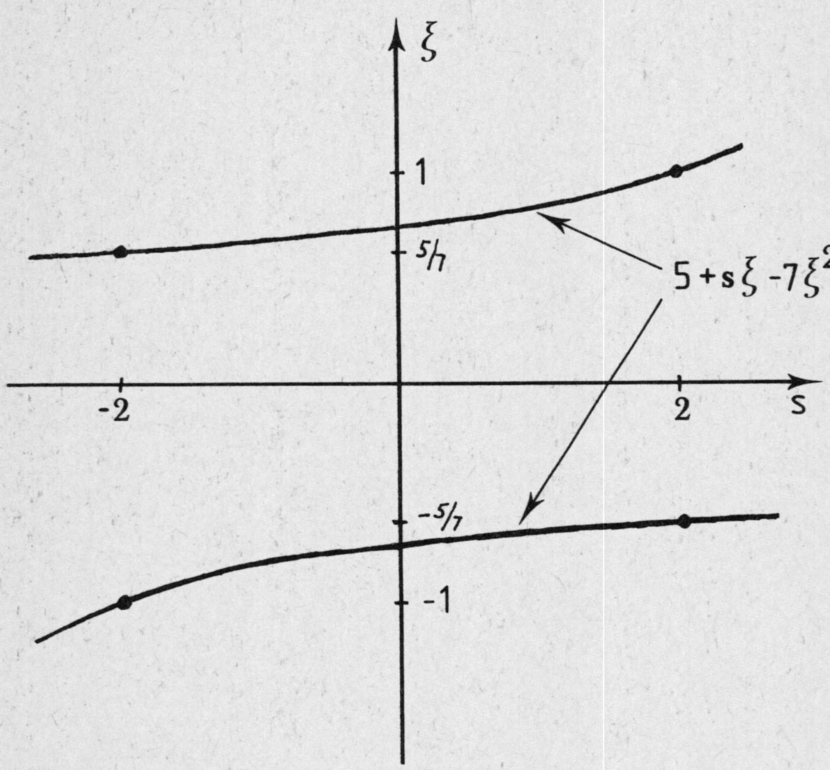

Figure 2. Graph of $5 + s\xi - 7\xi^2 = 0$ in [-2,2].

it follows that on (-2,2):

$$\text{if } -5/7 < \xi < 0, \text{ then } \xi'(s) = \frac{d\xi}{ds} < 0; \qquad (7)_a$$

or

if $\xi < -1$, then $\xi'(s) > 0$. $(7)_b$

From (6) and the continuity of ξ on $(-2,2)$, we deduce that ξ is continuously differentiable on $(-2,2)$.

The following argument shows that on $[-2,2]$, the range of ξ is $[-1,-5/7]$. Suppose that for some $s_0 \in (-2,2)$, $\xi(s_0) < -1$. Because $\xi(-2) = -1$, there is an $\tilde{s} \in (-2,s_0)$ such that $\xi'(\tilde{s}) < 0$. But this contradicts $(7)_b$. Therefore, $\xi \geqslant -1$ on $(-2,2)$. Next, suppose that $\xi(s_0) > -5/7$ for some $s_0 \in (-2,2)$. Since $\xi(2) = -5/7$ and ξ is continuous on $(-2,2)$, we may suppose that s_0 is chosen so that $0 > \xi(s_0) > -5/7$. Because ξ is continuous on $(-2,2)$, there is an $s_1 \in (-2,s_0)$ such that $\xi(s_1) = -5/7$. Consequently for some $\tilde{s} \in (s_0,s_1)$, $\xi'(\tilde{s}) > 0$. But this contradicts $(7)_a$. Therefore $\xi \leqslant -5/7$ on $(-2,2)$. Since $\xi(-2) = -1$ and $\xi(2) = -5/7$, we conclude that the range of ξ on $[-2,2]$ is $[-1,-5/7]$.

We now are in position to show that $\xi' > 0$ on $(-2,2)$. Suppose that $s_0 \in (-2,2)$ is an extremum of ξ. Then differentiating (6) with respect to s gives

$$6s\frac{d\xi}{ds} + 3(s^2-4)\frac{d^2\xi}{ds^2} = s\frac{d\xi}{ds} + \xi - 14\xi\frac{d\xi}{ds}$$

which evaluated at s_0 gives

$$3(s_0^2-4)\frac{d^2\xi}{ds^2}(s_0) = \xi(s_0) \qquad (8)$$

since $\frac{d\xi}{ds}(s_0) = 0$. Because $\xi(s) \in (-1,-5/7)$ when $s \in (-2,2)$, $\frac{d^2\xi}{ds^2}(s_0) > 0$. Therefore every extremum s_0 of ξ on $(-2,2)$ is a nondegenerate relative minimum. Since $\xi(-2) = -1 < \xi(s_0)$, there is an $s_1 \in (-2,s_0)$ where ξ has a relative maximum. This is a contradiction. Therefore ξ has no extrema on $(-2,2)$. Consequently ξ is strictly monotonic increasing on $[-2,2]$.

In §2 and §3 we will verify all the unproved assertions in the above argument.

§2. PICARD-FUCHS EQUATION

In this section we derive the Picard-Fuchs equation (1.5) satisfied by the periods of the differential forms ydx and xydx over the cycle γ_s on the Riemann surface of Γ: $y^2 = x^3 - 3x + s$.

Since we are interested in integrals over closed cycles, we will do all of our computations modulo exact forms. Differentiating

$$y^2 = x^3 - 3x + s \qquad (1)$$

with respect to x gives

$$2y\frac{dy}{dx} = 3(x^2-1) \qquad (2)$$

and with respect to s gives

$$2y\frac{dy}{ds} = 1. \qquad (3)$$

We compute

$$\begin{aligned} \alpha = ydx = -x\frac{dy}{dx}dx &= -\frac{3}{2}\frac{x}{y}(x^2-1)dx \qquad \text{using (2)} \\ &= -\frac{3}{2y}(y^2+2x-s)dx \qquad \text{using (1)} \\ &= -\frac{3}{2}\alpha - \frac{3}{2y}(2x-s)dx. \qquad (4) \end{aligned}$$

Integrating (4) over γ_s gives

$$\mathcal{A} = \frac{3s}{5}A - \frac{6}{5}B \qquad (5)$$

where

$$\mathcal{A} = \int_{\gamma_s} \alpha, \quad A = \int_{\gamma_s} \frac{dx}{y}, \text{ and } B = \int_{\gamma_s} \frac{x}{y}dx.$$

Also

$$\begin{aligned} \beta = xydx &= -\frac{1}{2}x^2\frac{dy}{dx}dx \\ &= -\frac{3x^2}{4y^2}(x^2-1)dx \qquad \text{using (2)} \\ &= -\frac{3x}{4y}(y^2+2x-s)dx \qquad \text{using (1)} \\ &= -\frac{3}{4}\beta - \frac{3}{4y}(2x^2-sx)dx, \end{aligned}$$

which implies

$$\begin{aligned} \beta &= -\frac{3}{7y}[2(1+\frac{2y}{3}\frac{dy}{dx})-sx]dx \qquad \text{using (2)} \\ &= -\frac{6}{7y}dx + \frac{3s}{7}\frac{x}{y}dx. \qquad (6) \end{aligned}$$

Integrating (6) over γ_s gives

$$\mathcal{B} = -\frac{6}{7}A + \frac{3s}{7}B \qquad (7)$$

where

$$\mathcal{B} = \int_{\gamma_s} \beta.$$

This completes the verification of (1.3). To find the Picard-Fuchs equation, we differentiate A and B with respect to s. Since integration over γ_s commutes with differentiation with respect to s we get

$$\frac{dA}{ds} = \int_{\gamma_s} \frac{dy}{ds}dx = \frac{1}{2}\int_{\gamma_s} \frac{1}{y}dx \quad \text{using (3)}$$

$$= \frac{1}{2}A, \qquad (8)$$

and

$$\frac{dB}{ds} = \int_{\gamma_s} x\frac{dy}{ds}dx = \frac{1}{2}\int_{\gamma_s} \frac{x}{y}dx$$

$$= \frac{1}{2}B. \qquad (9)$$

From (5) and (6) we get

$$\begin{pmatrix} A \\ B \end{pmatrix} = \frac{1}{3(s^2-4)} \begin{pmatrix} 5s & 14 \\ 10 & 7s \end{pmatrix} \begin{pmatrix} A \\ B \end{pmatrix}.$$

Therefore from (8) and (9) we obtain the Picard-Fuchs equation

$$\frac{d}{ds}\begin{pmatrix} A \\ B \end{pmatrix} = \frac{1}{6(s^2-4)} \begin{pmatrix} 5s & 14 \\ 10 & 7s \end{pmatrix} \begin{pmatrix} A \\ B \end{pmatrix}. \qquad (10)$$

§3. THE RICCATI EQUATION AND BOUNDARY VALUES

Here we derive the Riccati equation (1.5) satisfied by $\xi = B/A$. We also find the boundary values $\xi(-2)$ and $\xi(2)$.

Differentiating ξ and using (2.10) gives

$$3(s^2-4)\frac{d\xi}{ds} = 3(s^2-4)\left[\frac{1}{A}\frac{dB}{ds} - \frac{B}{A}\frac{1}{A}\frac{dA}{ds}\right]$$

$$= \frac{1}{2}\left[(10+7s\frac{B}{A}) - \frac{B}{A}(5s+14\frac{B}{A})\right]$$

$$= 5 + s\xi - 7\xi^2 \qquad (1)$$

which is the desired Riccati equation.

We now compute the boundary values of ξ. We begin with $\xi(2)$. At $s = 2$, the affine cubic Γ is reducible; in fact,

$$y^2 = x^3 - 3x + 2 = (x-1)^2(x+2). \qquad (2)$$

From

$$A(2) = \int_{\gamma_2} y dx = 2\int_{-2}^{1} (x-1)(x+2)^{\frac{1}{2}}dx = -\frac{8\cdot 3^{3/2}}{5} \qquad (3)$$

and

$$B(2) = \int_{\gamma_2} xy\,dx = 2\int_{-2}^{1} x(x-1)(x+2)^{\frac{1}{2}}dx = \frac{8\cdot 3^{1/2}}{7} \qquad (4)$$

we obtain $\xi(2) = -5/7$. Next we find $\xi(-2)$. Suppose that we establish the assertion:

> for every open set V in the x-y plane which contains $(-1,0)$, there is an $s_0 > -2$ with the following property: **for every $s\in(-2,s_0)$, the disc D_s, which is bounded by γ_s, is contained in V.**

Then Stokes' theorem together with a continuity argument will establish the value of ξ at $s = -2$. Before giving the details, we note that the above assertion expresses the geometrically obvious fact (which is easily seen from Figure 1) that, as s approaches -2 from above, the disc D_s, bounded by the compact component γ_s of the curve Γ, shrinks to the point $(-1,0)$. Assuming that the assertion holds, we compute $\xi(-2)$ as follows. Since the function $f(x,y) = x + 1$ is continuous at $(-1,0)$, for every $\varepsilon > 0$ there is an open set V containing $(-1,0)$ such that $|f(V)| < \varepsilon$. Thus, by the assertion, there is an $s_0 > -2$ such that for all $s \in (-2,s_0)$ $D_s \subseteq V$. Hence

$$\left|\int_{D_s} f(x,y)\,dxdy\right| \leqslant \varepsilon \int_{D_s} dxdy$$

that is, as $s \to -2^+$

$$\int_{D_s} (x+1)\,dxdy \Big/ \int_{D_s} dxdy \to 0. \qquad (5)$$

Now we compute $\xi(-2)$:

$$\xi(-2) = \lim_{s\to -2^+} \int_{\gamma_s} xy\,dx \Big/ \int_{\gamma_s} y\,dx$$

$$= \lim_{s\to -2^+} \int_{D_s} x\,dxdy \Big/ \int_{D_s} dxdy \qquad \text{by Stokes' theorem}$$

$$= \lim_{s\to -2^+} -\int_{D_s} dxdy \Big/ \int_{D_s} dxdy \qquad \text{using (5)}$$

$$= -1.$$

To complete the argument we need only verify the assertion. This we do using the Morse lemma. A straightforward calculation shows that $(-1,0)$

is a nondegenerate minimum of the function $H(x,y) = y^2 - x^3 + 3x$ where $H(-1,0) = -2$. Moreover the Hessian of H at $(-1,0)$ is $\begin{pmatrix} 6 & 0 \\ 0 & 1 \end{pmatrix}$. By definition we have

$$\gamma_s = H^{-1}(s) \cap \{x \leqslant 0\} \text{ and } D_s = (H^{-1}\{-2 < s\}) \cap \{x \leqslant 0\}.$$

From the Morse lemma, there is a local diffeomorphism $\varphi^{-1}(x,y) = (z,w)$ defined near $(-1,0)$ such that

$$\begin{aligned} H \circ \varphi(z,w) &= H(-1,0) + \tfrac{1}{2}D^2H(-1,0)((z,w),(z,w)) \\ &= -2 + \tfrac{1}{2}(6z^2+w^2). \end{aligned}$$

Let $\mathcal{D}_s = \{(z,w) \in \mathbb{R}^2 \mid \frac{1}{2}(6z^2+w^2) \leqslant s+2\}$. For every open neighborhood V of $(-1,0)$ where φ is defined, there is an $s_0 > -2$ such that for all $s \in (-2,s_0)$ $\mathcal{D}_s \subseteq \varphi^{-1}(V)$. Since $\varphi(\mathcal{D}_s) = D_s$, provided s_0 is close enough to -2, we have proved the assertion.

§4. A COROLLARY

We now prove the easy, but important

COROLLARY: *Consider the family of real affine cubic curves*

$$\Gamma: y^2 = P(x) + s$$

where P *is a real cubic polynomial. If for all* s *in a certain closed interval* I, Γ *has a compact component* γ_s, *then the function*

$$\xi(s) = \int_{\gamma_s} xydx \Big/ \int_{\gamma_s} ydx \qquad (2)$$

is strictly monotonic on I. *The same conclusion holds for the function*

$$\eta = \frac{a + b\xi}{c + d\xi} \qquad (3)$$

provided that $ad - bc \neq 0$.

PROOF: Using affine transformations (scaling and translation) Γ can be brought into the form

$$\tilde{\Gamma}: y^2 = x^3 \pm 3x + \tilde{s} \qquad (4)$$

where $\tilde{s}$ is a linear function of s. This affine transformation induces an affine transformation of ξ, that is,

$$\xi = e + f\tilde{\xi} \qquad (5)$$

where $\tilde{\xi}$ is the quotient of the period of xydx by the period of ydx over the cycle γ_s. Thus $\tilde{\xi}$ is strictly monotonic if and only if ξ is.

Since only those $\hat{\Gamma}$ with the minus sign have a compact component when s is in some interval, the corollary follows from the theorem.

Because

$$\frac{d\eta}{ds} = -\frac{(ad-bc)}{(c+d\xi)^2}\frac{d\xi}{ds},$$

η is strictly monotonic if ξ is strictly monotonic and $ad-bc \neq 0$. This completes the proof of the corollary.

§5. GLOBAL HOPF BIFURCATION

This section is based on Keener [2].

Consider his equation (2.5)

$$\frac{d\alpha}{d\tau} = \beta + \mu c_1\alpha + \varepsilon(\mu d_1\beta + \gamma d_2 + c_6\alpha^2) + \text{h.o.t.}$$
$$\frac{d\beta}{d\tau} = \lambda c_3\alpha + \gamma c_4 + c_5\alpha^2 + \mu c_2\beta + \varepsilon(d_3 + \lambda d_4)\beta + \text{h.o.t.} \qquad (1)$$

where ε and μ are small. Putting

$$x = \alpha$$
$$y = \frac{d\alpha}{d\tau}$$

and discarding the higher order terms together with terms in $\varepsilon^2, \varepsilon\mu$, and μ^2 gives

$$\frac{dx}{d\tau} = y \qquad (2)$$
$$\frac{dy}{d\tau} = \gamma c_4 + \lambda c_3 x + c_5 x^2 + \varepsilon(d_3 + \lambda d_4 + 2c_6 x)y + \mu(c_1 + c_2).$$

Let

$$H(x,y) = \tfrac{1}{2}y^2 - (\gamma c_4 + \tfrac{1}{2}\lambda c_3 x + \tfrac{1}{3}c_5 x^2)x = h. \qquad (3)$$

Then

$$\frac{dH}{dx} = \varepsilon(d_3 + \lambda d_4 + 2c_6 x)y + \mu(c_1 + c_2)y. \qquad (4)$$

The averaged equation coming from (4) is a linear combination of the integrals

$$\int_{\gamma_h} y\, dx \text{ and } \int_{\gamma_h} xy\, dx \qquad (5)$$

where γ_h is a compact component of the level set $H^{-1}(h)$. In what follows we will assume that $c_5 \neq 0$. To bring the family of affine cubics

$$\Gamma : \tfrac{1}{2}y^2 = (\gamma c_4+\tfrac{1}{2}\lambda c_3 x+\tfrac{1}{3}c_5 x^2)x + h \qquad (6)$$

into standard form, let

$$x_1 = x + \frac{\lambda c_3}{2c_5} \quad .$$

Then Γ becomes

$$\tilde{\Gamma} : y^2 = \tfrac{2}{5}c_5 x_1^3 - (\frac{\lambda^2 c_3^2}{2c_5} - 2\gamma c_4)x_1 + \frac{2}{3}\,\frac{\lambda^3 c_3^3}{4c_5^2} - \lambda\gamma\frac{c_3 c_4}{c_5} + h \qquad (7)$$

and (4) becomes

$$\frac{dH}{dx_1} = \varepsilon(d_3+\lambda d_4-\lambda\frac{c_3c_6}{c_5}+2c_6 x)y + \mu(c_1+c_2)y. \qquad (8)$$

When

$$\Delta = \lambda^2 c_4^2 - 4\gamma c_4 c_5 > 0 \qquad (9)$$

there is a nonempty closed interval I such that $\tilde{\Gamma}$ has a compact component when $h \in I$. From now on we assume that (9) holds. For $h \in I$, $\tilde{\Gamma}$ can be brought into Il'yashenko's normal form by scaling x_1 and y. Since this scaling does not affect monotonicity, we deduce that

$$\xi(h) = \int_{\gamma_h} xydx \Big/ \int_{\gamma_h} ydx$$

is strictly monotonic on I. Thus

$$\int_{\gamma_h} \frac{dH}{dx_1}\, dx_1 = 0$$

has a unique solution for all $h \in I$ when (ε,μ) lie in the cone C defined by

$$\varepsilon(d_3+\lambda d_4-\lambda\frac{c_3c_6}{c_5}+2c_6\xi) + (c_1+c_2)\mu = 0.$$

The exact range of ξ can be easily found once we have determined the scaling factors which bring (7) into Il'yashenko's normal form. Let

$$\begin{aligned} y &= \rho y_2 \\ x_1 &= \sigma x_2 \end{aligned} \qquad (10)$$

Then

$$\sigma = \Delta/2c_5$$

$$\rho = \Delta^{3/2}/2\sqrt{3}c_5$$

brings (7) into the desired normal form. Under the scaling defined by (10), ξ becomes

$$\xi_2 = \int x_2 y_2 dx_2 \Big/ \int y_2 dx_2 = \frac{1}{\sigma}\,\xi. \qquad (11)$$

Since the range of ξ_2 is $[-1,-5/7]$, the range of ξ is $[-\sigma,-5/7\sigma]$. Thus we have completely determined the cone C in the (ε,μ) plane for which there is a Hopf limit cycle for (2).

References

1. Il'yashenko, Yu.S., Zeros of special Abelian integrals in a real domain, Funct. Anal. and Appl. 11 (1977), 309-311.

2. Keener, J.P., Infinite period bifurcation: and global bifurcation branches, SIAM J. Appl. Math. 41 (1981), 127-144.

3. Il'yashenko, Yu.S., The multiplicity of limit cycles arising from perturbations of the form $w' = P_2/Q_1$ of a Hamiltonian equation in the real and complex domain, Trudy Sem. Petrovsk 3 (1978), 49-60 = AMS Transl. 118 (1982), 191-202.

4. Griffiths, P. and Harris, J., Principles of Algebraic Geometry, J. Wiley & Sons, New York, 1978.

5. Rauch, H. and Lebowitz, A., Elliptic functions, Theta functions and Riemann surfaces, Williams and Wilkins, Baltimore, Maryland, 1973.

6. Cushman, R. and Sanders, J., A codimension two bifurcation with third order Picard-Fuchs equation (to appear in J. Diff. Eqns.)

ON THE NUMERICAL DETERMINATION OF THE DIMENSION OF AN ATTRACTOR

by

Floris Takens

1. Introduction.

In [1] and independently in [4], a method was described to estimate a "dimension", or more precisely the correlation dimension, from experimental data. For a description of this method see also section 2 of the present paper. Various people asked me about the "convergence" of this algorithm, in other words about the errors to be expected in these estimates when only using a finite part of the data sequence. In this paper I discuss these errors and also extend this algorithm somewhat in order to minimize these errors.

In this discussion we use ideas from Mathematical Statistics, in particular Fischer's maximum likelihood rule, e.g. see [5]. In order to keep the paper self contained, I give complete proofs of the facts needed.

2. Description of the method.

The method of determining the correlation dimension of an attractor or ω-limit set in a physical experiment consists of two parts. First the "reconstruction of the phase space" in $\mathbb{R}^n$ and second the determination of the correlation dimension of a bounded subset in $\mathbb{R}^n$ with a probability measure. In the first part, a sequence $\{a_o,a_1,a_2,\dots\}$, obtained as a time sequence of experimental data ($a_i \in \mathbb{R}$), is used to produce a sequence $\{A_o,A_1,A_2,\dots\}$ in $\mathbb{R}^n$; this sequence is defined by $A_i = (a_i,a_{i+1},\dots,a_{i+n-1})$. We assume that this sequence $\{A_i\}_{i\in\mathbb{N}}$ is bounded in $\mathbb{R}^n$ and define its limit set A by $A = \{x\in\mathbb{R}^n | x$ is an accumulation point of $\{A_i\}\}$. We also assume that $\{A_i\}$ defines a probability measure on A: for $K\subset A$ closed, and K_ε the ε-neighbourhood of K in $\mathbb{R}^n$, we define the measure of K as

$$m(K) = \lim_{\varepsilon\to o} \left(\lim_{n\to\infty} \frac{1}{N} \cdot (\#\{0 \leq i \leq N | A_i \in K_\varepsilon\})\right)$$

(the assumption that this measure exists implies the existence of these limits). The experiment admits a finite dimensional deterministic description (or model) with phase space X, if the sequence $\{a_i\}$ is (in this model) obtained by composing the orbit $x(i)\in X$, describing the evolution during the experiment (with discrete time $i\in\mathbb{N}$), with a "read out" map $y: X\to\mathbb{R}$ (so $a_i = y(x(i))$). Then the sequence $\{A_i\}$ and the set A in $\mathbb{R}^n$ are the image of the orbit $\{x(i)\}$ and it ω-limit under a smooth transformation of X into $\mathbb{R}^n$. In [3] we proved that if n is sufficiently big, and if y, togehter with the dynamics in X satisfy certain generic conditions, then this transformation from X to $\mathbb{R}^n$ is in fact an embedding. So then A is a faithfull representation of the ω-limit (or attractor).

If the collection of experimental data has started after the transient phenomena (i.e. phenomena due to particular initial conditions) have died out, then we may assume that the points A_i belong to A. This we assume from now on.

The correlation dimension for such a set $A \subset \mathbb{R}^n$ with probability measure m is defined by the following formulae, using the product measure m^2 on $A \times A$:

$$\text{correlation dimension } (A) = \lim_{\varepsilon \to 0} \frac{\ln P(\varepsilon)}{\ln \varepsilon}$$

with $P(\varepsilon) = m^2(\{(a_1,a_2) \in A \times A \mid \rho(a_1,a_2) \leq \varepsilon\})$; ρ refers to the Euclidean distance in $\mathbb{R}^n$. If $\lim_{\varepsilon \to 0}$ does not exist, the correlation dimension is not defined. This notion was introduced independently in [1] and [4]. For a numerical determination of the correlation dimension one generates a large number of (random) pairs (A_i,A_j) (from the sequence $\{A_i\}$ defining A and m), for example all pairs (A_i,A_j) with $0 \leq i < j \leq N$ for some N, and counts the number of pairs for which the distance is smaller than ε for various values of ε. In this way one obtains estimates for P(ε), for these values of ε. In many cases, e.g. see [2], it then turns out that the estimated values of ln(P(ε)) depend, to a good approximation, linearly on ln ε. The slope of ln(P(ε)), as function of ln ε, determines the correlation dimension. In this way it is not easy to estimate the accuracy of this method. Especially not because, for smaller ε, the estimate of ln(P(ε)) becomes more inaccurate.

For this reason we propose here a somewhat different way to proceed. This will lead to an estimate for the correlation dimension for which we know the standard error and for which this standard error is minimal in a sense to be discussed in section 4. In this procedure we also start with a (finite) sequence of distances $\rho_1,\rho_2,\ldots$ of (randomly) choosen pairs (A_i,A_j). Assuming that the limit, defining the correlation dimension, exists, we have $P(\varepsilon) = c.\varepsilon^\alpha$ + higher order terms (where α is the correlation dimension). We now take some $\varepsilon_0 > 0$ fixed, and assume that for $\varepsilon \leq \varepsilon_0$ we have exactly $P(\varepsilon) = c.\varepsilon^\alpha$; we then try to estimate α.

This means that we disregard all distances ρ_i which are bigger than ε_0 and use the remaining distances to estimate α. This problem, of estimating α from these distances, can be formulated independently of ε_0 by dividing all distances by ε_0. In this way we get the following problem:

let $r_1,r_2,\ldots,r_m \in [0,1]$ be a random sample from a probability distribution (so we use R for the random variabele and $r_1,\ldots,r_m$ as the values of this variabele in particular experiments). What is the most likely value of α?

In the next section we deal with this problem.

In section 5 we discuss some cases where the distances ρ_i, as far as they are smaller than ε_0, are not distributed according to $P(\varepsilon) = c.\varepsilon^\alpha$.

3. The maximum likelihood rule.

According to Fischer's maximum likelihood rule, e.g. see [5], we should proceed as follows. If we find a sample $r_1,\dots,r_m$ (see the end of section 2) then we observe that the probability of finding a sample in $(r_1, r_1+dr_1),\dots,(r_m, r_m+dr_m)$, dependent on α is $\Pi\alpha.r^{\alpha-1}dr_i$. We calculate the value of α for which this is maximal: taking the logarithm of the last expression gives $m.\ln\alpha+(\alpha-1).\Sigma\ln r_i$ which attains its maximum for $\frac{m}{\alpha}+\Sigma\ln r_i=0$ or for $\alpha^{-1}=\overline{\ln r_i}$ ($\overline{\ln r_i}$ indicates the average value of $\ln r_i$, $i=1,\dots,m$). This means that $(-\overline{\ln r_i})^{-1}$ is the most likely value of α, given the sample $r_1,\dots,r_m$. This is at least a simple rule to estimate α; in what sense it is "most likely" we shall discuss in section 4.
In this section we only derive the standard error of this estimate.

First we calculate, for a given distribution $\alpha.R^{\alpha-1}.dR$ on $[0,1]$ the mean- or expectation value of $\ln R$:

$$\mathcal{E}(\ln R)=\int_0^1 \ln r.\alpha.r^{\alpha-1}\,dr = r^\alpha \ln r\,\Big|_0^1 - \int_0^1 \frac{r^\alpha}{r}.\,dr = -\frac{1}{\alpha}.$$ For this calculation one has to assume that $\alpha\neq 0$, but the case where $\alpha=0$, or α is near zero, is the one giving the least problems in the numerical determination of the correlation dimension. From now on we assume that the parameter α is positive. For the expectation value of $(\ln R)^2$ we find:

$$\mathcal{E}((\ln R)^2)=\int_0^1(\ln r)^2.\alpha.r^{\alpha-1}dr = r^\alpha(\ln r)^2\,\Big|_0^1 - \int_0^1\frac{2\ln r}{r}.r^\alpha dr = -2\int_0^1 \ln r.r^{\alpha-1}dr = \frac{2}{\alpha^2}.$$

This means that the variance of $\ln R$ equals $\sigma(\ln R)=\mathcal{E}((\ln R)^2)-(\mathcal{E}(\ln R))^2=\frac{1}{\alpha^2}$.
So the standard error when estimating the average of $\ln R$, or $\mathcal{E}(\ln R) = -\frac{1}{\alpha}$, from a sample of size m is

$$\frac{1}{\sqrt{m}}.\frac{1}{\alpha}.$$

For m large, this is small. So, for large m we can obtain the standard error in the estimate of α by multiplying the above quantity with the absolute value of the derivative of the function $f(x)=-\frac{1}{x}$ (see the figure below). In this way we obtain $\frac{1}{\sqrt{m}}.\alpha$.

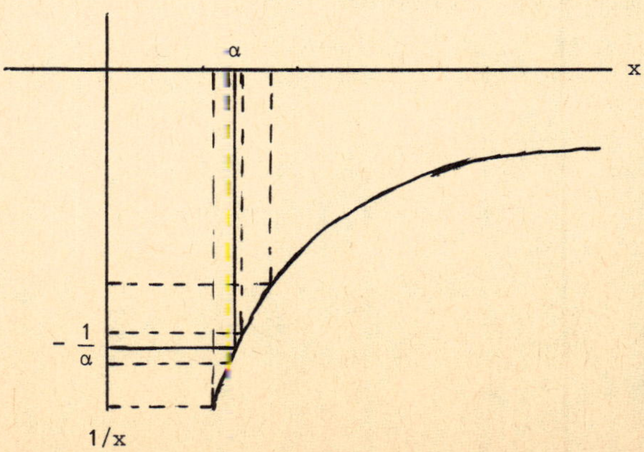

So for example , the estimate of the correlation dimension α will have a standard error of 10% if the sample size is 100.

The above considerations were carried out for known and fixed α. However if we only know the sample $r_1,\dots,r_m$ they can also be used: we first estimate α and then use the estimated value of α to determine the standard error of the estimate.

Finally we observe that, especially in case of small sample sizes, the estimate of α will have a tendency of being in average somewhat high. For the estimate of $-\frac{1}{\alpha}$ or $\mathcal{E}(\ln R)$ this is not the case, but due to the non-linearity of $f(x) = -\frac{1}{x}$ a symmetric interval around $-\frac{1}{\alpha}$ gets transformed into a non-symmetrical interval around α, see the above figure.

4. Optimality.

We come back to the problem at the end of section 2. On $[0,1]$ we have a probability distribution $\alpha.R^{\alpha-1}.dR$, with α unknown. We have to estimate α on the basis of a random sample $r_1,\dots,r_m$. In the previous section we saw how the maximum likelihood rule suggested to estimate α by taking the inverse of the average of $-\ln r_i$. Here we consider a larger class of procedures, namely we consider estimates of α, based on taking the average of $g(r_i)$ for some function $g\colon [0,1] \longrightarrow \mathbb{R}$ (of course we have to restrict to those functions for which the expectation value $\mathcal{E}_\alpha(G) = \int_0^1 g(r).\alpha.r^{\alpha-1}.dr$ exists for all $\alpha > 0$) writing G for $g(R)$. We shall show that we get the best results when taking $g(R) = \ln(R)$. We need to explain what we mean here by "best". For this we calculate the standard error when estimating α on the basis of the average of $g(r_i)$ when m, the sample size, is large.

We denote by $\mathcal{E}_\alpha(G)$, resp. $\mathcal{E}_\alpha(G^2)$, the expectation value of G, resp. G^2, with respect to the probability distrubution $\alpha.r^{\alpha-1}.dr$. We compute, just as in the case with $g(r) = \ln(r)$, the standard error when estimating $\mathcal{E}_\alpha(g)$ from a sample of size m. This gives

$$\frac{1}{\sqrt{m}}.\sqrt{\mathcal{E}_{\bar\alpha}(G^2) - (\mathcal{E}_{\bar\alpha}(G))^2}\text{ , when the actual value of } \alpha \text{ is } \bar\alpha.$$

For large m this is small (we assume that $\mathcal{E}_{\bar\alpha}(G^2)$ exists). In order to obtain an estimate for α, we have to use the inverse of the function of

$$\alpha \longmapsto \mathcal{E}_\alpha(G).$$

If this inverse does not exist, then we cannot estimate α from the average of $g(r_i)$ but then also g will not be acceptable for the criterion we give below. So from now on we assume that $\alpha \longmapsto \mathcal{E}_\alpha(G)$ is invertable. Again as in the case where $g(r) = \ln(r)$, we get the standard error, when estimating α, by multiplying the standard error of $\mathcal{E}_\alpha(G)$, see above, by the absolute value of the inverse of the derivative of $(\alpha \longmapsto \mathcal{E}_\alpha(G))$ in $\alpha = \bar\alpha$:

$$\frac{1}{\sqrt{m}} \cdot \frac{\sqrt{\mathcal{E}_{\overline{\alpha}}(G^2) - (\mathcal{E}_{\overline{\alpha}}(G))^2}}{|\partial_\alpha \mathcal{E}_{\overline{\alpha}}(G)|} .$$

This we want to be as small as possible for α and m fixed. So we define the <u>quality of g</u>, as function of $\overline{\alpha}$, by

$$Q(g,\overline{\alpha}) = \frac{(\partial_\alpha(\mathcal{E}_{\overline{\alpha}}(G)))^2}{\mathcal{E}_{\overline{\alpha}}(G^2) - \mathcal{E}_{\overline{\alpha}}(G))^2} .$$

So the <u>main problem</u> is to find those g, depending on $\overline{\alpha}$, for which this quantity attains its maximum value (for $\overline{\alpha}$ fixed). In fact we shall show that this optimal function g is independent of $\overline{\alpha}$ and equals $g(R) = \ln(R)$.

In order to prove this we calculate $Q(g,\overline{\alpha})$ for $g(r) = \ln(r) + \tilde{g}(r)$. We first show that it is no loss of generality to assume that

$$* \int_o^1 \tilde{g}(r).\overline{\alpha}.r^{\overline{\alpha}-1}.dr = 0 \text{ and } \int_o^1 \ln(r).\tilde{g}(r).\overline{\alpha}.r^{\overline{\alpha}-1}dr = 0.$$

This follows from the fact that the quality of a function g does not change if we add a constant to g or if we multiply g with a non-zero scalar. So we may replace g(r) by $\lambda_1(\ln(r) + \tilde{g}(r)) + \lambda_2 = \ln(r) + (\lambda_1 - 1).\ln r + \lambda_1.\tilde{g}(r) + \lambda_2$, i.e., we may replace $\tilde{g}(r)$ by $(\lambda_1 - 1)\ln r + \lambda_1.\tilde{g}(r) + \lambda_2$. By ajusting λ_1 and λ_2, we can get the above qualities except if $\tilde{g}$ satisfies the following quation:

$$-\frac{1}{\overline{\alpha}^2} - \frac{1}{\overline{\alpha}}.\mathcal{E}_{\overline{\alpha}}(\tilde{G}) - \mathcal{E}_{\overline{\alpha}}(\ln R.\tilde{G}) = 0 \qquad (1).$$

So if we prove that for all $\tilde{g}$ not satisfying (1) we have $Q(\ln + \tilde{g},\overline{\alpha}) \leq Q(\ln r,\overline{\alpha})$, then by continuity we have the result also for those $\tilde{g}$ satisfying (1).

Now we calculate $Q(\ln + \tilde{g}, \overline{c})$ using the above equalities $*$. To follow these calculations it is useful to remember that

$$\int_o^1 r^\beta.\ln r.dr = -\frac{1}{(\beta+1)^2} , \int_o^1 r^3.(\ln r)^2.dr = \frac{2}{(\beta+1)^3} , \beta > -1.$$

$$\mathcal{E}_{\overline{\alpha}}(G) = \int_o^1 (\ln r + \tilde{g}(r)).\overline{\alpha}.r^{\overline{\alpha}-1}dr = -\frac{1}{\overline{\alpha}}$$

$$\partial_\alpha(\mathcal{E}_\alpha(G)) = \int_o^1 \partial_\alpha((\ln r + \tilde{g}(\alpha)\cdot\alpha.r^{\alpha-1})dr = \int_o^1 (\ln r + \tilde{g}(r))(r^{\alpha-1} + \alpha.r^{\alpha-1}.\ln r)dr.$$

$$\text{So } \partial_\alpha(\mathcal{E}_{\overline{\alpha}}(G)) = \int_o^1 \ln r(r^{\overline{\alpha}-1} + \overline{\alpha}.r^{\overline{\alpha}-1}.\ln r)dr = \frac{1}{\overline{\alpha}^2} .$$

$$\mathcal{E}_{\overline{\alpha}}(G^2) = \int_0^1 (\ln r + \tilde{g}(r))^2.\overline{\alpha}.r^{\overline{\alpha}-1}dr = \int_0^1 (\ln r)^2.\overline{\alpha}.r^{\overline{\alpha}-1} + \int_0^1 (\tilde{g}(r))^2.\overline{\alpha}.r^{\overline{\alpha}-1}dr \geqq \frac{2}{\overline{\alpha}^2} ;$$

the equality only holds if $\tilde{g}(r) \equiv 0$.

From this it follows that if $\tilde{g} \not\equiv 0$ and if $\tilde{g}$ satisfies $*$, then $Q(g,\overline{\alpha}) < Q(\ln, \overline{\alpha})$. This solves our main problem.

In the language of mathematical statistics, our results can be stated as follows: $\overline{-\ell n(r_i)}$ is an unbiased estimate for $\frac{1}{\alpha}$ with minimal variance (this can also be derived from [5,§38]); $\overline{-\ell n(r_i)}^{-1}$ is an assymptotically unbiased estimate for α with assymptotically minimal variance.

5. Concluding remarks.

In the preceeding two sections we assumed that we were dealing with a probability distribution $\alpha.r^{\alpha-1}.dr$ on $[0,1]$ (corresponding to a probability distribution $c.\rho^{\alpha-1}.d\rho$ on $[0,\varepsilon_o]$, or corresponding to $P(\varepsilon) = c'.\varepsilon^{\alpha}$ for $\varepsilon \leqq \varepsilon_o$). This assumption is not always justified. In fact, in a number of cases one has observed that for some ε_1, $P(\varepsilon)$ is approximately proportional to $c_1.\varepsilon^{\alpha_1}$ for $\varepsilon < \varepsilon_1$ and is approximately proportional to $c_2.\varepsilon^{\alpha_2}$ for $\varepsilon > \varepsilon_1$. This means that the "observed" or estimated dimension may depend on the scale (ε_o) of our observations. We calculate the result of such a probability distribution when applying our method of averaging $-\ln(\rho_i)$.

For convenience we transform again the interval $[0,\varepsilon_o]$ to $[0,1]$, replacing the variable $\rho \in [0,\varepsilon_o]$ by $r \in [0,1]$. If $P(r)$ denotes the probability of r' to belong to $[0,r]$, then we want

$$P(r) = \begin{cases} c_1.r^{\alpha_1} & \text{for} \quad r \leqq r_o \\ c_2.r^{\alpha_2} & \text{for} \quad r \geqq r_o ; \qquad 0 < r_o < 1. \end{cases}$$

Since $P(1) = 1$, $c_2 = 1$. Since the two definitions of $P(r_o)$ should agree, $c_1 = r_o^{\alpha_2-\alpha_1}$. For this probability distribution we calculate the expectation value of $\ln(r)$:

$$\mathcal{E}(\ln(R)) = \int_0^1 \ln r.dP(r) = \int_0^{r_o} \ln r.c_1.\alpha_1.r^{\alpha_1-1}\, dr + \int_{r_o}^1 \ln r.\alpha_2.r^{\alpha_2-1}.dr =$$

$$c_1\{\ln r.r^{\alpha_1}\Big|_0^{r_o} - \int_0^{r_o} r^{\alpha_1-1}.dr\} + \{\ln r.r^{\alpha_2}\Big|_{r_o}^1 - \int_{r_o}^1 r^{\alpha_2-1}.dr\} =$$

$$r_o^{\alpha_2-\alpha_1}\{r_o^{\alpha_1}.\ln r_o - \frac{r_o^{\alpha_1-1}}{\alpha_1}\} + \{-r_o^{\alpha_2}.\ln r_o - \frac{1}{\alpha_2}(1-r_o^{\alpha_2})\} = -\frac{1}{\alpha_2} + r_o^{\alpha_2}(-\frac{1}{\alpha_1} + \frac{1}{\alpha_2}).$$

Some consequences of this formula are:

- In the case where these is a "random noise" in the signal $\{a_1,a_2,\ldots\}$ due to experimental (or round off) errors we may expect a probability distribution for r of the above type with $\alpha_1 = \infty$ and r_1, or rather $\varepsilon_o . r_1$ (see the transformation from ρ to r above) or the order of the experimental errors. In this case the random noise gives a correction term $\frac{1}{\alpha_2} . r_o^{\alpha_2}$. It is interesting to observe that this term decreases rapidly as α_2 takes higher values. So for higher dimensional attractors the effect of experimental errors becomes smaller.
- If α_1 is very near zero, $\mathcal{E}(\ln(R))$ will be very strongly negative (leading to estimates of the correlation dimension near zero). So in this case our way of estimating the correlation dimension tends to ignore the α_2 part. Now however we cannot say much about the standard error in the determination of $\mathcal{E}(\ln R)$.

In any case the above calculation shows that it is important to verify whether the distribution of distances in $[0,\varepsilon_o]$ really satisfies $P(\varepsilon) \sim c.\varepsilon^{\alpha}$. This can be done by repeating the estimation of the correlation dimension for various values of ε_o.

There is even an example of Zaslavski [2] [6] where numerical results indicate that $P(\varepsilon) \sim c.\varepsilon^{\alpha}$ does not hold on any interval of the form $[0,\varepsilon_o]$; with the present algorithm I found the estimates of the correlation dimension fluctuate from ± 1.1 to ± 1.5.

On the other hand there is an example (namely the interval mapping $x \longmapsto 1 - \mu x^2$ on $[-1,+1]$ for μ just below the Feigenbaum point 1.4011...) where we know that we should find the correlation dimension to be zero, but where we find, except for very low values of ε_o ($< 10^{-2}$), either an estimate for the correlation dimension around 0.5, or we find that some r_i is zero (leading to ln 0).

Finally some remarks on how to choose ε_o. Of course ε_o must be bigger than the experimental error; the ratio between ε_o and the experimental error should be bigger if one deals with lower dimensional attractors. Also ε_o should be sufficiently big so that the sample of distances $\rho \leq \varepsilon_o$ is not to small; however this sample size can be raised by considering more pairs (A_i,A_j).
Also ε_o should not be too big: A is bounded so $P(\varepsilon) \sim c.\varepsilon^{\alpha}$ will certainly be violated for large values of ε. Especially when the correlation dimension is big, the last two requirements may force one to consider very large numbers of pairs (A_i,A_j).

References.

1. P. Grassberger, I. Procaccia, On the characterization of strange attractors, Phys.Rev.Lett. 50 (1983), 346-349.

2. ———————, Measuring the strangeness of strange attractors, Physica 9 D (1983), 189-208.

3. F. Takens, Detecting strange attractors in turbulence, in Dynamical systems and turbulence, Warwick 1980, Lecture Notes in Mathematics, 898, Springer-Verlag, Berlin.

4. ———————, Invariants related to dimension and entropy, in Atas do 13° colóqkio brasileiro de matemática, Rio de Janeiro, 1983.

5. B.L. van der Waerden, Mathematical Statistics, Grundlehren vol. 156, Springer Verlag, Berlin.

6. G.M. Zaslavskii, The simplest case of a strange attractor, Phys.Lett. 69 A (1978), 145-147.

F. Takens
Mathematisch Instituut
Postbus 800
9700 AV Groningen
The Netherlands

Global stability of generic two-parameter families of gradients on three-manifolds

Gert Vegter

Introduction

One of the earliest results in the theory of differentiable dynamical systems was the fact that generic gradient vector fields on a closed n-manifold are structurally stable. It was obtained as a by-product of the more general work of Palis and Smale [6], in which structural stability of Morse-Smale diffeomorphisms was established.

The Morse-Smale gradient vector fields constitute an open and dense subset G of the set $\mathcal{X}^g$ (M) of all gradient systems on a closed manifold M. Systems in G are characterized by the following features:

a) all singularities are hyperbolic (i.e. non-degenerate)

b) all stable and unstable manifolds are transversal.

A gradient system is structurally stable if and only if it belongs to G. Palis and Smale even proved that a system in G is not merely topologically equivalent, but even conjugated to any sufficiently nearby system in G.

A common approach in Bifurcation Theory (often called the "Thom-Smale program") is to allow instabilities by having exactly one of the generic conditions a) and b) violated in the weakest possible sense. This situation will be met in generic one-parameter families of gradients at isolated values of the parameter. After introduction of a parametrized version of the concept of structural stability, the question arises whether these one-parameter families are generically structurally stable.

This problem was settled by Palis and Takens [7]. Imposing several additional generic conditions on these families, they proved structural stability of generic one-parameter gradient families on closed n-manifolds.

In this situation there are obstructions ("moduli") to the existence of a topological conjugacy between two sufficiently close gradient families near an orbit of tangency between stable and unstable manifolds (see [3],[4]). However, once a smooth reparametrization of time has been carried out, this obstruction is no longer present.

The possibility of constructing a topological conjugacy, in stead of merely a topological equivalence, between a family of gradients and a slightly perturbed, possibly reparametrized family has the advantage that one may use <u>invariant foliations</u>. These will be described further on in detail. They are rather crucial in the proof of the afore mentioned results, especially when the dimension of the manifold M exceeds 2.

Continuing our way in the Thom-Smale program we consider situations where the violation of conditions a) and b) is two-fold, or where exactly one of these conditions is violated, while simultaneously one of the additional conditions, imposed by Palis and Takens, does not hold. These situations are met in generic two-parameter families of gradients at isolated values of the parameters.

Our main result is:

Theorem. There is an open and dense subset G of the set $\mathcal{X}_2^g(M^3)$ of two-parameter families of gradients on a closed 3-manifold, such that each family in G is structurally stable.

In the situation, described in the theorem, it is quite hard to find a topological conjugacy between two sufficiently close members of G, even up to a smooth reparametrization of time. It might even be impossible in general. Consequently it is hard or even impossible to use invariant foliations to the same extent as in the presence of at most one parameter. Instead of these we shall introduce singular foliations on some parts of the manifold.

Our methods apply to gradient families on closed 3-manifolds, but it is not obvious how to extend them to useful tools in higher dimensions. Problems arise in connection with the occurrence of a degenerate orbit of tangency between invariant manifolds. Here we require some invariant foliations to be smooth, which leads to serious technical difficulties if the number of dimensions exceeds three, cf. [10].

Subsequently one might pose the question whether in general k-parameter families of gradients are structurally stable, as Palis and Takens did in [7].

However, since the methods employed are rather complicated and of a quite ad-hoc nature, proving this conjecture in general seemed to be a hopeless enterprise. Fortunately, a recent counter-example of Takens [8] revealed that is is even an impossible one: structural stability is not a generic property of k-parameter gradient families on n-manifolds if $k \geq 8$ and $n \geq 3$. If one takes $k \geq 16$ and $n \geq 4$ there are even infinitely many "moduli".

There is obviously a gap between the positive results of this chapter and these counterexamples. This leads to:

Problem

Prove or disprove:

Structural stability is not a generic property of k-parameter families of gradients on closed n-manifolds for $k \geq 3$ and $n \geq 3$.

Conjecture

Generic k-parameter families of gradients on closed 2-manifolds are structurally stable.

Outline of the paper

First we consider generic one- and two-parameter families of gradients on two-manifolds. The bifurcations occurring in these families are rather simple. Similar bifurcations may be met in higher dimensions.
Subsequently we consider some tools that are useful for dealing with matters of structural stability in higher dimensions. The use of various kinds of invariant foliations is demonstrated by reproving the well known theorem of Hartman-Grobman on local structural stability for gradients near a hyperbolic singular point. Extension of these methods to global situations is indicated briefly. The main problem turns out to be the bifurcation associated with the occurrence of an orbit of tangency between invariant manifolds. We give necessary and sufficient conditions for stability of one parameter families of gradients on three-manifolds exhibiting this kind of bifurcation. We also indicate why singular foliations are useful tools for the attack of this situation.
Finally we describe the bifurcations of two-parameter families on three-manifolds corresponding to the occurrence of exactly one orbit of tangency which exhibits some kind of degeneracy. These bifurcations do not occur in gradient families on two-manifolds.
The paper is intended to be rather self contained, so some parts may be quite familiar to the reader.

Acknowledgements

I am indebted to Henk Broer and Floris Takens for carefully reading the manuscript and for suggesting some improvements.

Basic features of gradients

A gradient vector field $X=\text{grad}_g f$ on a compact C^∞ manifold M is associated with a C^∞ Riemannian metric g and a C^∞ function $f:M\to\mathbb{R}$ by the relation

$$g(X,Y) = df(Y) \qquad (1)$$

for all smooth vector fields Y on M. From (1) it is obvious that f is strictly increasing along regular orbits of X, since the potential f is a Lyapunov function for X outside singular points of f. So $\text{grad}_g f$ has no periodic orbits or other kinds of recurrency. The singular points of X are exactly those of f. Since the linear part LX of $X = \text{grad}_g f$ at a singularity is symmetric with respect to the inner product defined by g, all eigenvalues of LX are real (in suitable coordinates LX is the Hessian matrix of f at the singular point).

One parameter families of gradients on a two-manifold

In this section we shall describe the instabilities that may occur generically, even open- en densely, in two parameter families of gradients at isolated values of the parameter. We first restrict our attention to systems in dimension two. These bifurcations are also met in higher dimensions, but in these cases our study can be confined to an invariant manifold of dimension two in which the situation is not essentially different from that to be described presently. This reduction of the number of dimensions is well known in the study of singular points: there the invariant manifold is a center manifold. In a later section we shall deal with unstable gradients which do not admit a similar reduction of the number of dimensions.

A k-parameter family of gradients $\{X^\mu\}$ on a (compact) manifold M is a family of pairs $(g^\mu,f^\mu),\mu\in\mathbb{R}^k$, where $\{g^\mu\}$ is a k-parameter family of Riemannian metrics and $\{f^\mu\}$ is a k-parameter family of real-valued functions on M, such that $X^\mu = \mathrm{grad}_{g^\mu} f^\mu$.
We assume that both g^μ and f^μ, and hence X^μ, depends C^∞ on $(\mu,x) \in \mathbb{R}^k x M$.
The set of such pairs is endowed with the strong Whitney topology, i.e. the topology of uniform convergence of g^μ, f^μ and all their derivatives on compact sets. The resulting topological space of k-parameter families is denoted by $\mathcal{X}^g_k(M)$. A property P is said to hold open- and densely (generically) for k-parameter families of gradients if there is an open dense (residual) subset of $\mathcal{X}^g_k(M)$ such that each member of this subset has property P.
As was pointed out in the introduction we pay special attention to the property of being structurally stable. Before we give a precise setting for this study we shall describe the instabilities occurring open- and densely in two-parameter families of gradients on two-manifolds.
A parameter value $\bar\mu \in \mathbb{R}^k$ is called a bifurcation value for a family $\{X^\mu\}$ in $\mathcal{X}^g_k(M)$ if $X^{\bar\mu}$ fails to be a Morse-Smale system. In virtue of the characterization of Morse-Smale gradients (see introduction) $X^{\bar\mu}$ has at least one orbit of tangency between stable and unstable manifolds or at least one non-hyperbolic singular point.
The bifurcation set of a family $\{X^\mu\}$ in $\mathcal{X}^g_k(M)$ is the subset of $\mathbb{R}^k$ consisting of all bifurcation values of $\{X^\mu\}$.

For an open and dense set of one-parameter families of gradients it is easy to describe the bifurcation diagram: it consists of isolated points in the parameter space $\mathbb{R}$ at which exactly one of the conditions appearing in the characterization of Morse-Smale gradients is violated. Figure 1 contains a pictorial representation of the bifurcation diagram and of the changing dynamics of the corresponding gradients.

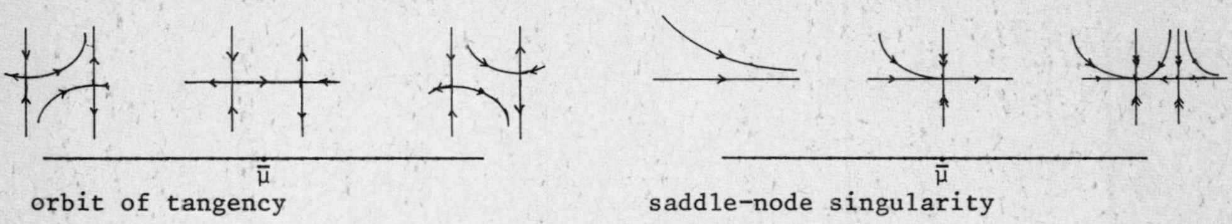

Figure 1

These scenarios only hold for families $\{X^{\mu}\}$ belonging to an open and dense subset of $\mathcal{X}_1^g(M)$: the instability of $X^{\bar{\mu}}$ should "unfold generically". For the orbit of tangency this means that the corresponding invariant manifolds should cross at non-zero speed as the parameter passes the value $\bar{\mu}$. To be more precize, consider a one-dimensional transversal section Σ on the orbit of tangency, as indicated in figure 1. The (signed) distance of the invariant manifolds, measured along Σ grows as const. $(\mu-\bar{\mu})+O((\mu-\bar{\mu})^2)$ for μ near $\bar{\mu}$. For the saddle-node it boils down to the creation (annihilation) at $\mu=\bar{\mu}$ of a pair of singular points, whose distance grows (shrinks) as constant. $\sqrt{|\mu-\bar{\mu}|}+O(|\mu-\bar{\mu}|)$ for $\mu>\bar{\mu}$ $(\mu<\bar{\mu})$. A more precize description is contained in [7] and [10].

Remark

In dimension two an orbit of tangency between invariant manifolds is just a saddle connection.

Generic two-parameter families of gradients

The instabilities described above may also occur open- and densely in two-parameter gradient families on M^2. The corresponding parameter values then constitute a smooth curve in parameter space $\mathbb{R}^2$. Moreover at an isolated value $\bar{\mu}$ of the parameter exactly one of the following situations may occur:

I $X^{\bar{\mu}}$ has exactly two orbits of tangency (saddle connections in dimension two) between stable and unstable manifolds, while all singularities are hyperbolic.

II $X^{\bar{\mu}}$ has exactly one non-hyperbolic singularity, which is a saddle-node, and exactly one orbit of tangency between invariant manifolds.

III $X^{\bar{\mu}}$ has exactly two non-hyperbolic singularities, which are saddle-nodes, while all invariant manifolds intersect transversally.

IV $X^{\bar{\mu}}$ has exactly one non-hyperbolic singularity, which is quasi-hyperbolic of type 2 (see Remark 1), while all invariant manifolds are transversal.

Remarks

1. Although $X^{\bar{\mu}}$ is supposed to be a gradient system on a two-manifold, the description above is also valid in higher dimensions.

2. A point $s \in M$, (M not necessarily of dimension two) is called a quasi-hyperbolic singularity of type 2 of a vector field X if there is a one dimensional, locally invariant manifold W^c (center manifold) of class C^m, such that:
   a) 0 is an eigenvalue of the linear part of X at s of multiplicity one, while all other eigenvalues are off the imaginary axis;
   b) on W^c there is a local C^m-coordinate x, such that $X|W^c = x^3.X_1(x)\frac{\partial}{\partial x}$, with $X_1(0) \neq 0$.

   W^c is not unique in general. The degree of differentiability m may be taken arbitrarily large, see [2].
   In this context a saddle-node might be called a quasi-hyperbolic singularity of type 1.

3. If $X^{\bar{\mu}}$ has a non-hyperbolic singular point $s \in M$ (again M not necessarily of dimension two), as in case II - IV, the set of invariant manifolds of $X^{\bar{\mu}}$ also includes the strong (un-)stable manifolds of this singular point, i.e. the boundaries of the (un-)stable manifolds of s. These manifolds, denoted by $W^{ss}(s)$ and $W^{uu}(s)$, are smooth and uniquely determined by the fact that $T_sW^{ss}(s)$ and $T_sW^{uu}(s)$ are the generalized eigenspaces corresponding to the eigenvalues with negative and positive real parts, respectively (see Figure 2; also see [2]).

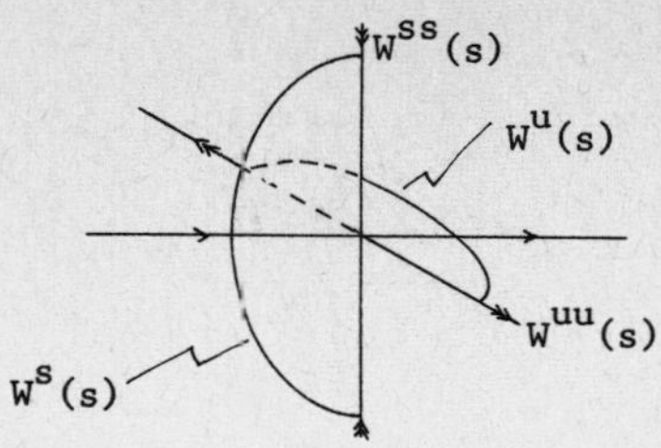

Figure 2

Next we consider the bifurcation sets in $\mathbb{R}^2$ near parameter values $\bar{\mu}$ for which we have one of the situations described above. Such a parameter value is in the closure of smooth curves in $\mathbb{R}^2$, which correpond to the occurrence of bifurcations that also may occur in one-parameter families. We shall briefly describe these bifurcations and refer to Figure 3 for a more complete impression of the changing dynamics of X^μ, for μ near $\bar{\mu}$.

case I.

Here there are two essentially different cases:

a. There are four hyperbolic singular points s_i^μ $(i=1,..,4)$ of $X^{\bar{\mu}}$ such that the orbits of tangency are contained in $W^u(s_1^{\bar{\mu}}) \cap W^s(s_2^{\bar{\mu}})$ and $W^u(s_3^{\bar{\mu}}) \cap W^s(s_4^{\bar{\mu}})$, respectively, see Figure 3.1.a.
b. There are three hyperbolic singular points $s_i^{\bar{\mu}}(i=1,..,3)$ of $X^{\bar{\mu}}$ such that the orbits of tangency are contained in $W^u(s_1^{\bar{\mu}}) \cap W^s(s_2^{\bar{\mu}})$ and $W^u(s_1^{\bar{\mu}}) \cap W^s(s_3^{\bar{\mu}})$, respectively.

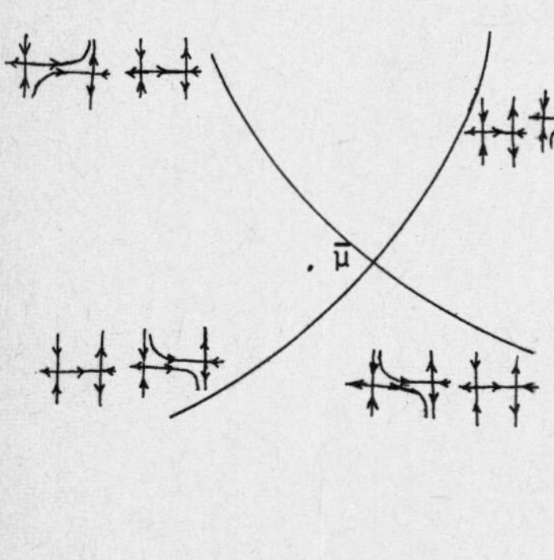

Figure 3.1.a.

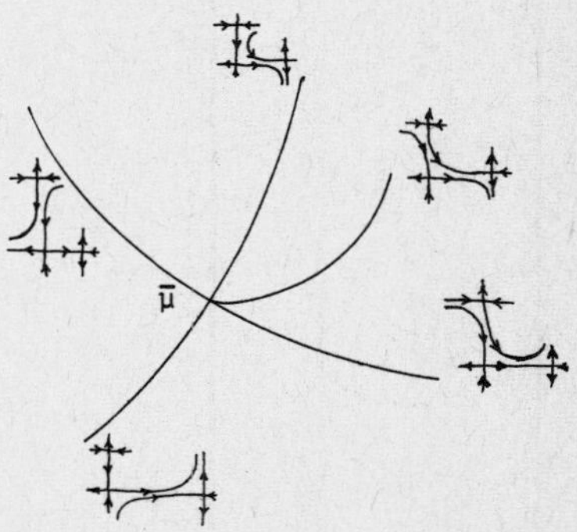

Figure 3.1.b.

case II.

We distinguish the following situations:

a. $X^{\bar{\mu}}$ has one saddle-node $s^{\bar{\mu}}$ and one orbit of tangency between $W^u(s_1^{\bar{\mu}})$ and $W^s(s_2^{\bar{\mu}})$, where $s_1^{\bar{\mu}}$ and $s_2^{\bar{\mu}}$ are hyperbolic saddle-points of $X^{\bar{\mu}}$ (fig.3.2.a.).

b. $X^{\bar{\mu}}$ has a saddle-node $s^{\bar{\mu}}$ and a hyperbolic saddle-point $\tilde{s}^{\bar{\mu}}$ whose stable manifold has one orbit of tangency with the unstable manifold of $\tilde{s}^{\bar{\mu}}$. Secondary bifurcations are, among others, due to the occurrence of tangencies between $W^u(s_*^{\mu})$ and $W^s(\tilde{s}^{\mu})$, for each saddle-point $s_*^{\mu} \neq \tilde{s}^{\mu}$ for which $W^u(s_*^{\bar{\mu}}) \cap W^s\ (s^{\bar{\mu}}) \neq \emptyset$, see figure 3.2.b.

c. $X^{\bar{\mu}}$ has a saddle-node $s^{\bar{\mu}}$ and an orbit of tangency between $W^{uu}(s^{\bar{\mu}})$ and $W^s(\tilde{s}^{\bar{\mu}})$, where $\tilde{s}^{\bar{\mu}}$ is a saddle-point of $X^{\bar{\mu}}$

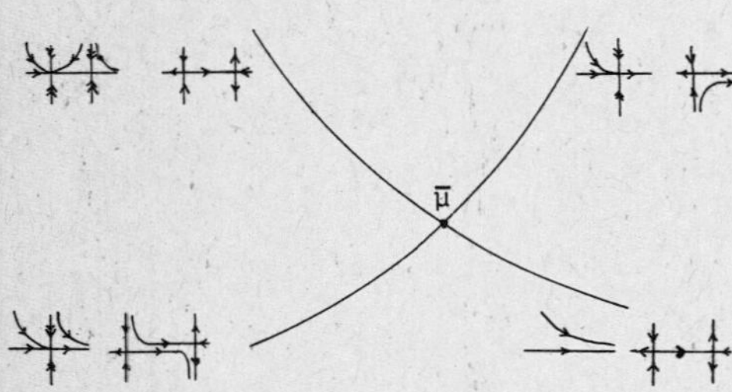

Figure 3.2.a.

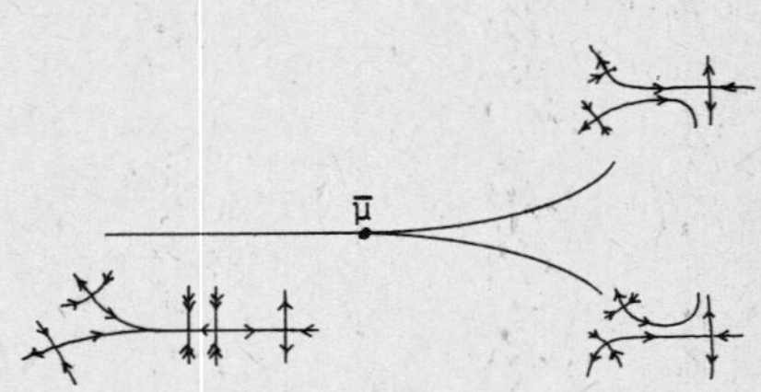

Figure 3.2.b.

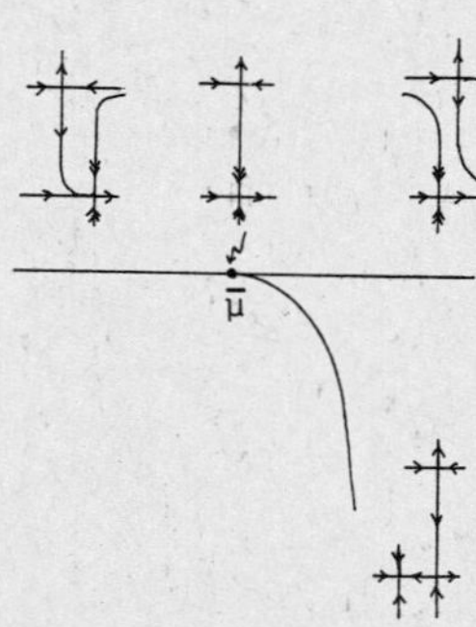

Figure 3.2.c

Remark

Saddle connections of type II.c were treated more generally in [9]. There we studied unfoldings of vector fields with a general non-hyperbolic singular point whose strong (un-)stable separatrix coincides with a separatrix of a hyperbolic saddle-point.

case III.

For an open and dense class of two parameter families $\{X^\mu\}$ of $\mathcal{X}_2^g(M)$ we have a pair s_1, s_2 of saddle-nodes occurring at isolated values $\bar{\mu}$ of the parameter. There are two curves Γ_1 and Γ_2 in the parameter plane corresponding to the occurrence of exactly one saddle-node of X^μ near s_1 and s_2, respectively. Generically these curves are transversal.

case IV.

In a neighbourhood of the central bifurcation value $\bar{\mu}$ the bifurcation diagram consists of parameter values μ for which X^μ, and hence f^μ, has a degenerate singular point near s. For an open and dense class of two parameter families $\{f^\mu\}$, for which grad f^μ has a quasi-hyperbolic singularity of type 2, there are μ-dependent local coordinators x,y in which f^μ can be written as: $\pm x^4 + \mu_1 x^2 + \mu_2 x \pm y^2$ (cf.[1]). So the bifurcation diagram near $\bar{\mu}$ is the well-known cusp, see Figure 3.3. The pair of curves having μ in their closure correspond to the occurrence of exactly one saddle node near s.

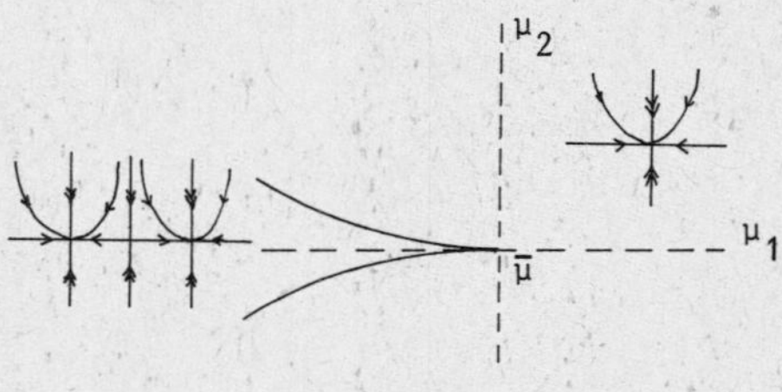

Figure 3.3

Remark.

The bifurcations we just described also occur in higher dimensions: in fact the description of the changing dynamics is not restricted to the two dimensional case. However in dimension three (and higher) there is an additional type of bifurcation corresponding to the occurrence of exactly one orbit of tangency accompanied by some kind of degeneracy. This situation will be treated in a later section.

Structural stability of families of gradients.

In this section we recall the notion of topological equivalence between vector fields and extend it to parametrized families. We investigate whether an open and dense class of these parametrized families has the property that the members are in the interior of their equivalence classes, i.e. they are structurally stable.

To be more precize we say that two vector fields X and X' on a manifold M are topologically equivalent if there is a homeomorphism $h: M \to M$ sending orbits of X to orbits of X' in a sense preserving way. A vector field X is structurally stable if any sufficiently near by (with respect to the strong Whitney topology on the set of all vector fields on M) vector field is topologically equivalent to X. As a first result in this context we mention:

Theorem (Grobman/Hartman)

If X is a vector field having a hyperbolic singular point at $p \in M$, then X and its linear part $(dX)_p$ are topologically equivalent on a neighbourhood U of p.

One may interpret this theorem as a result on structural stability: if Y is a vector field which is sufficiently near X, then Y has a (unique) hyperbolic singular point q near p and there are neighbourhoods U of p and V of q such that $X|U$ and $Y|V$ are topologically equivalent. This corollary of the theorem is an immediate consequence of the fact that hyperbolic linear vector fields are structurally stable (within their own world).

As mentioned in the introduction, the structurally stable gradient vector fields form an open and dense subset G of $\mathcal{X}^g(M)$, characterized by the following features:

a) all singularities are hyperbolic
b) all stable and unstable manifolds are transversal.

The study of parametrized families requires an extension of the concept of structural stability:

Two k-parameter families of gradients $\{X^\mu\}$ and $\{\hat{X}^\mu\}$ are called topologically equivalent at $\bar{\mu}$ and $\hat{\mu}$ if there is a neighbourhood U of $\bar{\mu}$ in $\mathbb{R}^k$ and a continuous mapping $H: U \times M \to \mathbb{R}^k \times M$, which is a homeomorphism onto its image, such that:

a) H is a of the form $H(\mu,x) = (h(\mu),H^{\mu}(x))$, where $h: (U,\bar{\mu}) \to (\mathbb{R}^k,\hat{\mu})$ is a homeomorphism onto its image.

b) $H^{\mu}: M \to M$ is a topological equivalence between X^{μ} and $\hat{X}^{h(\mu)}$.

The corresponding concept of stability is:
A k-parameter family of gradients $\{X^{\mu}\} \in \mathcal{X}^g_k(M)$ is <u>structurally stable at $\bar{\mu} \in \mathbb{R}^k$</u> if there is a neighbourhood U of $\bar{\mu}$ in $\mathbb{R}^k$ and a neighbourhood $\mathcal{U}$ of $\{X^{\mu}\}$ in $\mathcal{X}^g_k(M)$ such that for each family $\{X^{\mu}\} \in \mathcal{U}$ there is a $\hat{\mu} \in U$ such that $\{X^{\mu}\}$ and $\{\hat{X}^{\mu}\}$ are topologically equivalent at $\bar{\mu}$ and $\hat{\mu}$.

Palis and Takens [7] proved that structural stability holds for members of an open and dense subset G_1 of $\mathcal{X}^g_1(M)$. At isolated values of the parameter families in G_1 exhibit bifurcations due to violation of exactly one of the conditions a) and b) that determine the set of G of structurally stable gradients.
The class G_1 is characterized by a number of conditions which hold open and densely. However they are not necessary for structural stability of a one parameter family. We shall state necessary conditions further on and indicate how the theorem of Palis and Takens can be proved for this slightly broader class of gradient families. For this purpose we have to introduce invariant foliations that exhibit a rather mild kind of singularity.

<u>Remark</u>.
It is desirable to have necessary conditions for structural stability of k-parameter families, since under these circumstances bifurcations occurring in (k+1)-parameter families at isolated values of the parameter involve violation of exactly one of these conditions. Violation of a sufficient condition does not necessarily give rise to a new type of bifurcation when adding more parameters.

<u>Invariant foliations</u>.
We now give an outline of a proof of the theorem of Hartman and Grobman, since it is a nice occasion to introduce the concept of invariant foliations (also cf. [4]). The situation described in the theorem is depicted in Figure 4.
In the unstable manifold of p we take a fundamental domain D^s, i.e. a sphere with $\dim D^s = \dim W^u(p) - 1$ which is transversal to X. Over D^s we raise a small cylinder C^s which is also transversal to X. Next we take a regular foliation $\mathcal{F}_c$ of C^s whose leaves are of the same dimension as $W^s(p)$, and transversal to D^s. The stable <u>foliation $\mathcal{F}^s(p)$</u> is now obtained by saturation of the foliation $\mathcal{F}_c$ and by adding $W^s(p)$ as a special leaf, i.e. a leaf F of $\mathcal{F}^s(p)$, different

from $W^s(p)$, is of the form $X_t(F')$, where F' is a leaf of $\mathcal{F}_c$ and t is a real number. By construction the foliation $\mathcal{F}^s(p)$ is invariant for the flow $\{X_t\}_{t\in\mathbb{R}}$ of X. As a consequence of the λ-lemma (see [4]) the foliation $\mathcal{F}^s(p)$ is regular in the sense that its leaves constitute a continuous family of C^1-embedded discs.

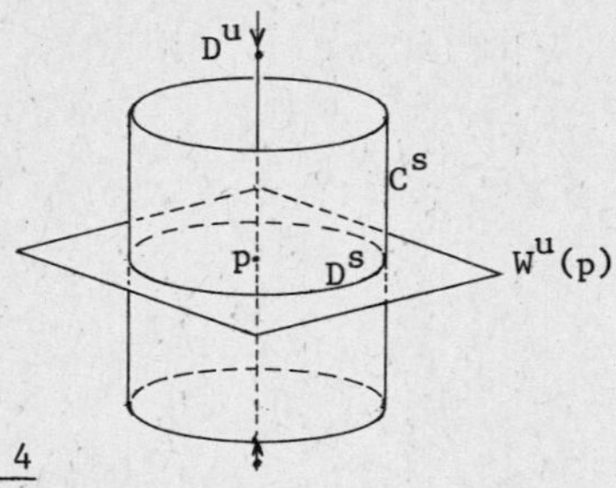

Figure 4

In the same way we obtain an invariant unstable foliation $\mathcal{F}^u(p)$, using a fundamental domain D^u contained in $W^s(p)$. Note that the space of leaves of $\mathcal{F}^u(p)$ and $\mathcal{F}^s(p)$ is represented by $W^s(p)$ and $W^u(p)$, respectively. Together these foliations may be used as coordinates on a neighbourhood of p.
A topological equivalence between X and dX_p near p may be obtained by introducing a similar pair $\tilde{\mathcal{F}}^u(p)$, $\tilde{\mathcal{F}}^s(p)$ of invariant foliations for dX_p, and by fixing the induced homeomorphisms h^s and h^u on the spaces of leaves of $\mathcal{F}^s(p)$ and $\mathcal{F}^u(p)$, respectively. To fix h^s we take fundamental domains D^s and $\tilde{D}^s$ in the unstable manifolds of p for X and dX_p, respectively. Subsequently we define an arbitrary homeomorphism $h^s: D^s \to \tilde{D}^s$ and extend it to all of $W^u(p)$ by the requirement that (i) $h^s(p)=p$ and (ii) $X_t \circ h^s = h^s \circ (dX_p)_t$, for all $t \in \mathbb{R}$. In the same way we obtain h^u on the space of leaves of $\mathcal{F}^u(p)$, which settles the construction of the topological equivalence h.

An essential feature of the construction described above is the fact that we obtain a <u>topological conjugacy</u> between X and dX_p, i.e. $X_t \circ h = h \circ (dX_p)_t$, for all $t \in \mathbb{R}$. This follows from the construction of h^s and h^u, and from the fact that $\mathcal{F}^s(p)$ and $\mathcal{F}^u(p)$ are invariant for the flow of X. In some cases there are obstructions to the existence of such a conjugacy, see [5]. Sometimes there is a way out by first performing a reparametrization of time (as for instance in [7]), but in more awkward situations even this may be insufficient. To overcome these problems, at least in some situations, we introduce singular foliations. Their use will be illustrated by reproving the theorem of Hartman and Grobman in the case of a saddle-point of a gradient system in dimension three, see Figure 5.

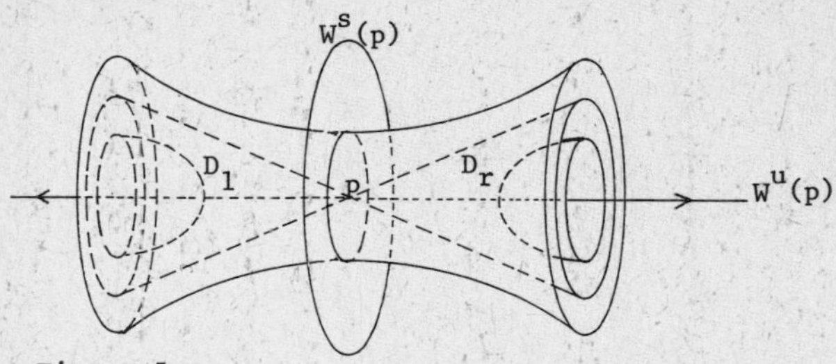

Figure 5

To obtain a topological equivalence h between $X=\text{grad}_g f$ and dX_p near p we require that h respects level surfaces of f. So we first fix the induced homeomorphism on the space of level surfaces.

Proceeding naively, we might try to define h by fixing it on a pair of discs D_l, D_r, each of which is contained in a regular level surface of f and both intersecting $W^u(p)$. Although in this way h would be defined at points near p, outside $W^s(p)$, it is in general not possible to extend h continuously to all of $W^s(p)$. To this very end we introduce a singular foliation $\mathcal{S}^u(p)$ near p. Again we take a cylinder C over a fundamental domain contained in $W^s(p)$ and introduce a regular foliation on C with leaves of dimension one. We take C in a level surface of f (see Figure 6). The leaves of $\mathcal{S}^u(p)$ are

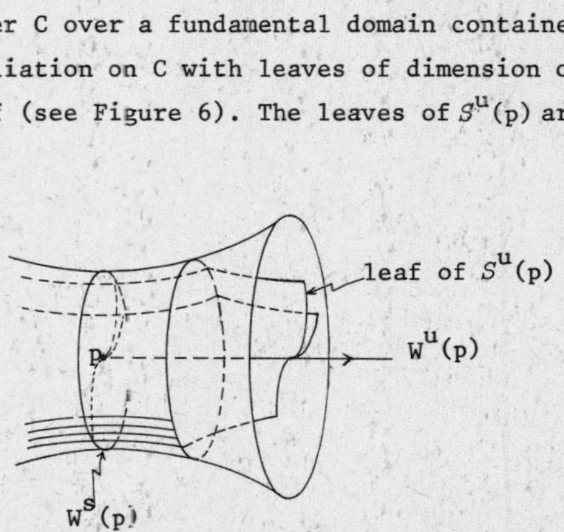

Figure 6

of the form $\bigcup_{t\geq 0} X_t(F)$, where F is a leaf of the foliation on C.

Introducing a similar foliation for dX_p we may now complete our construction of h by fixing the induced homeomorphism h_* on the space of leaves of $S^u(p)$, and finally defining h on the discs D_l and D_r in a way compatible with h_*.

Remarks.

1. Note that leaves of $S^u(p)$ consist of (parts of) orbits of X, in contrast with the invariant stable and unstable foliations described above. Apart from the exceptional leaves formed by the invariant manifolds the latter foliations have no leaves that are tangent to X.
2. If p is a hyperbolic saddle point of X such that $W^u(p)$ has dimension two we may define a singular foliation $S^s(p)$ in a similar fashion.
3. Suppose the contracting eigenvalues of dX_p are different. In that case there is a unique direction in T_qD_r $(q=W^u(p)\cap D_r)$, represented by some vector $v \in T_qD_r$ for which $\lim_{t\to-\infty} (dX_t)_q(v)$ is the strongest contracting eigendirection of dX_p, see figure 7.

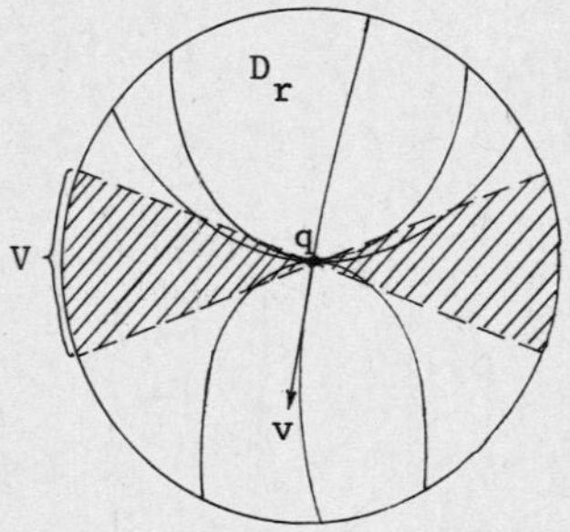

Figure 7

In this case it is sufficient for continuous extendibility of h to $W^s(p)$ that $h|D_r$ is compatible with the homeomorphism on the space of leaves of $S^u(p)$ only in a cone-like setV in D_r with vertex q. V does not contain the leaf of $S^u(p)\cap D_r$ tangent to v. Every other leaf intersects Vand is transversal to the boundary of V. In this way continuous extendibility of h to the one dimensional strong stable manifold $W^{ss}(p)\subset W^s(p)$ is guaranteed, since h also preserves level sets of f. This observation, which I owe Floris Takens, adds a crucial amount of freedom to the construction of h in some delicate situations (cf. [10]).

<u>Global constructions</u>.

Invariant foliations are especially suitable for the construction of global topological equivalences. To illustrate this consider the situation depicted in Figure 8. Here p and q are hyperbolic saddle points of a Morse-Smale gradient vector field X such that $W^u(p) \cap W^s(q) \neq \emptyset$.

If a homeomorphism $h: M \to M$ should be a global topological <u>conjugacy</u> between X and some vector field X' which sends leaves of the invariant foliations of X to leaves of the corresponding foliations of X' the foliations $\mathcal{F}^s(p)$ and $\mathcal{F}^s(q)$ should be <u>compatible</u>. By this we mean that if a leaf F of $\mathcal{F}^s(p)$ has non-empty intersection with a leaf F' of $\mathcal{F}^s(q)$, then F is contained in F'. To achieve compatibility we first fix $\mathcal{F}^s(q)$ as described before and subsequently we get the foliation of the cylinder C as an extension of the partial foliation formed by the intersection of C and leaves of $\mathcal{F}^s(q)$. In this way it is possible to associate compatible systems of stable and unstable foliations with each Morse-Smale gradient. These are used for the construction of a global topological conjugacy in the same way as described in the proof of the theorem of Hartman and Grobman. In our description we omitted many technical details which are rather involved. For a complete construction of compatible systems of invariant foliations we refer to [7].

<u>Remark</u>.

When using a singular foliation on a neigbourhood of p the construction of a topological equivalence will be of a hybrid nature: outside some neighbourhood U of p we get a topological conjugacy with the aid of invariant stable and unstable foliations, inside U it will be constructed using singular foliations. Compatibility of the singular foliation and the remaining invariant foliations will be achieved as above (also see [10]).

<u>Tangencies of stable and unstable manifolds in generic one parameter families of gradients</u>.

In this section we consider the case where $\{X^\mu\}$, $\mu \in \mathbb{R}$, is a generic one parameter family of gradients on a 3-manifold M, such that $X^{\bar\mu}$ $(\bar\mu \in \mathbb{R})$ has an orbit of non-transveral intersection of the unstable manifold $W^u(s_1^{\bar\mu})$ of $s_1^{\bar\mu}$ and the stable manifold $W^s(s_2^{\bar\mu})$ of $s_2^{\bar\mu}$.

The following condition is satisfied by generic one parameter families of gradients as above:

(i) For $r \in \gamma$ we have $\dim(T_r W^u(s_1^{\bar\mu}) + T_r W^s(s_2^{\bar\mu})) = 2$

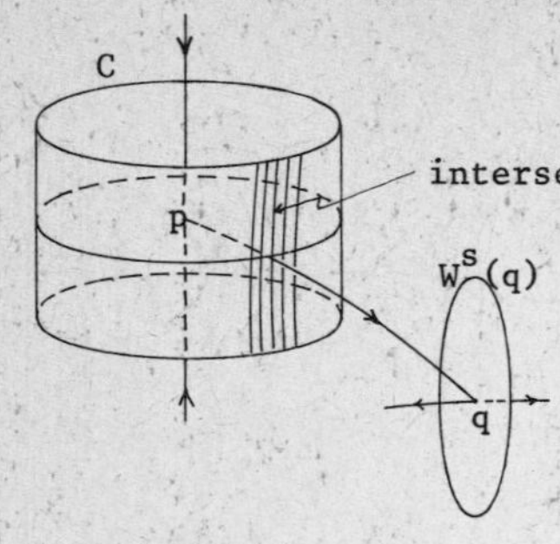

Figure 8

So generically we have three cases:

I. dim $W^u(s_1^{\bar{\mu}})$=dim $W^s(s_2^{\bar{\mu}})$=2.

II. dim $W^u(s_1^{\bar{\mu}})$=1, dim $W^s(s_2^{\bar{\mu}})$=2.

II*. dim $W^u(s_1^{\bar{\mu}})$=2, dim $W^s(s_2^{\bar{\mu}})$=1.

Case II* can be treated similar to case II by reversing the sign of X^μ, so we don't consider it separately.

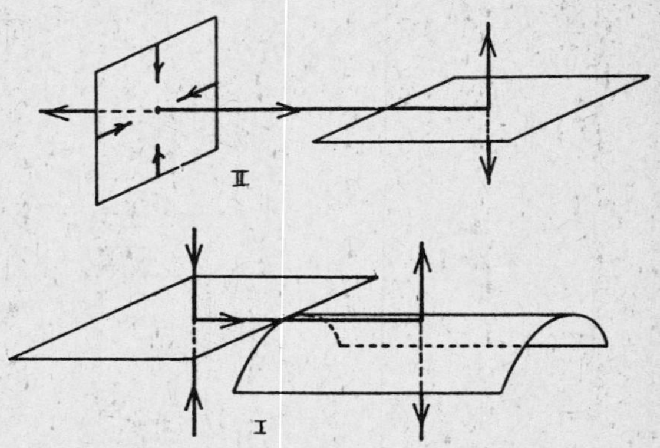

Figure 9

In case I we impose the condition that the tangency between $W^u(s_1^{\bar{\mu}})$ and $W^s(s_2^{\bar{\mu}})$ is as generic as possible in the following sense.
Let S be a smooth two dimensional cross-section for $X^{\bar{\mu}}$, containing some point $r \in \gamma$. Take coordinates x_1,x_2 on a neighbourhood of r in S, in which $W^u(s_1^{\mu}) \cap S = \{x_2 = 0\}$, while $W^s(s_2^{\mu}) \cap S$ is of the form $\{x_2 = F(x_1)\}$, for some smooth function F.
Condition (i) amounts to: $F(0) = \frac{dF}{dx}(0) = 0$.

The following condition is satisfied by generic one-parameter families for which (i) holds:
(ii) $\frac{d^2F}{dx_1^2}(0) \neq 0$.

Remark.
Condition (ii) doesn't depend on the choice of r,S and the local coordinates x_1,x_2. In fact it can be stated in an intrinsic way, which also makes sense in higher dimensions (cf. condition (Q.T.2) in [7]).
A tangency of this kind is called quasi-transversal.

The following condition expresses a "crossing-at-non-zero-speed" of $W^u(s_1^\mu)$ and $W^s(s_2^{\bar\mu})$ *) as the parameter passes the value $\bar\mu$. For this we choose σ^u, σ^s: $\mathbb{R}\to M$ with $\sigma^u(\mu) \in W^u(s_1^\mu)$, $\sigma^s(\mu) \in W^s(s_2^\mu)$ and $\sigma^u(\bar\mu) = \sigma^s(\bar\mu) = r \in \gamma$. We require:

(iii) $\quad \dot\sigma^u(\bar\mu) - \dot\sigma^s(\bar\mu) \neq 0 \mod(T_r W^u(s_1^{\bar\mu}) + T_r W^s(s_2^{\bar\mu}))$

Remark.
This condition is independent of the choice of σ^u and σ^s. Note that $\mu \mapsto s^\mu$ is a smooth mapping, associating with μ the unique singularity of X^μ, close to $s^{\bar\mu}$ (for μ close to $\bar\mu$). Also not that the invariant manifolds $W^u(s_1^\mu)$ and $W^s(s_2^\mu)$ depend smoothly on μ on compact parts.

If we have a tangency as in case II, we impose the following generic condition on $\{X^\mu\}$:

(iv) The contracting eigenvalues of the linear part of $X^{\bar\mu}$ at $s_1^{\bar\mu}$ are different (this is generic for gradients because only real eigenvalues can occur).

If in case II condition (iv) is satisfied there is a unique one dimensional invariant manifold $W^{ss}(s_1^{\bar\mu}) \subset W^s(s_1^{\bar\mu})$ such that $T_{s^{\bar\mu}}(W^{ss}(s_1^{\bar\mu}))$ is the eigenspace of the linear part of $X^{\bar\mu}$ at $s_1^{\bar\mu}$, corresponding to the strongest contracting eigenvalue. $W^{ss}(s_1^\mu)$ is called the strong stable manifold of $s_1^{\bar\mu}$.
Let $E_r \subset T_r M$ $(r \in \gamma)$ be a one dimensional subspace of $T_r W^s(s_2^{\bar\mu})$, complementary to $X^{\bar\mu}(r)$. We finally impose the following conditions:

(v) If $X^{\bar\mu}$ has an orbit of tangency as in case II, then $\lim_{t\to-\infty} (dX_t^{\bar\mu})_r(E_r) = T_{s^{\bar\mu}} W^{ss}(s_1^{\bar\mu})$ ($X_t^{\bar\mu}$, $t \in \mathbb{R}$, is the flow of $X^{\bar\mu}$).

(vi) If $X^{\bar\mu}$ has an orbit of tangency as in case II, then the stable and unstable manifolds of any singularity $s^* \neq s_1^{\bar\mu}, s_2^{\bar\mu}$, are transversal to $W^{ss}(s_1^{\bar\mu})$.

(vii) If s^* is a singularity of $X^{\bar\mu}$, are different from $s_1^{\bar\mu}$ and $s_2^{\bar\mu}$, then s^* is hyperbolic and the stable and unstable manifolds of s^* are transversal to the stable and unstable manifolds of any other singularity of $X^{\bar\mu}$.

*) $s_1^\mu(s_2^\mu)$ is the saddle-point of X^μ near $s_1^{\bar\mu}(s_2^{\bar\mu})$.

Conditions (i) - (vii) are necessary and sufficient for structural stability of the family $\{X^\mu\}$. The necessity of these conditions will become obvious from the description of the bifurcations of generic two-parameter families of gradients to be described in the next section. Here we briefly indicate why these conditions are sufficient.

Suppose $\{\tilde{X}^\mu\}$ is a one-parameter family of gradients near $\{X^\mu\}$, such that $\tilde{X}^{\tilde{\mu}}$ has an orbit of tangency between $W^u(\tilde{s}_1^{\tilde{\mu}})$ and $W^s(\tilde{s}_2^{\tilde{\mu}})$, satisfying (i) - (vii) for μ near $\tilde{\mu}$. We indicate how to obtain a topological equivalence H of the form $H(\mu,x) = (h(\mu), H^\mu(x))$ in case II (cf.fig.9). First the change of parameter h is taken in such a way that $h(\bar{\mu})=\tilde{\mu}$. We introduce μ-dependent singular foliations $\mathcal{S}^u(s_1^\mu)$ and $\mathcal{S}^u(\tilde{s}_1^\mu)$ near s_1^μ and $\tilde{s}_1^\mu$, respectively. Recall that these foliations are taken to be compatible with invariant foliations $\mathcal{F}^u(s_*^\mu)$ of hyperbolic saddle points s_*^μ for which $W^u(s_*^\mu) \cap W^s(s_1^\mu) \neq \emptyset$. In particular part of the invariant manifold of such saddle points is a leaf of $\mathcal{S}^u(s_1^\mu)$. For these saddle points we first fix the homeomorphisms on the spaces of leaves of $\mathcal{F}^u(s_*^\mu)$. Subsequently we take the homeomorphism H_*^μ on the space of leaves of $\mathcal{S}^u(s_1^\mu)$ in such a way that it is compatible with these homeomorphisms. To extend H^μ to a full neighbourhood of s_1^μ we take a continuous family of smooth discs D_r^μ, μ near $\bar{\mu}$, with D_r^μ contained in a level set of f^μ, such that $W^u(s_1^\mu) \cap D_r^\mu \neq \emptyset$. Similarly we introduce $\mathcal{S}^u(\tilde{s}_1^\mu)$ and $\tilde{D}_r^\mu$ for $\tilde{X}^\mu$ (cf. Figures 6 and 7). We fix H^μ on D_r^μ in such a way that it is compatible with H_*^μ on a cone-like neighbourhood S^μ of $r = W^u(s_1^\mu) \cap D_r^\mu$ in D_r^μ (see Figure 7 and the related remark). Moreover we take H^μ such that is sends $W^s(s^\mu) \cap D_r^\mu$ to $W^s(\tilde{s}^{h(\mu)}) \cap D_r^{h(\mu)}$. This is possible since these sets are transversal to leaves of the singular foliations <u>at points of S^μ</u>, as a consequence of condition (v). As in the proof of the theorem of Hartman and Grobman it is now easy to extend H^μ continuously to a full neighbourhood of s_1^μ, and subsequently to a neighbourhood of s_2^μ.

In case I the construction of H^μ is less involved. We take a one parameter family of cylinders C^μ over a fundamental domain in $W^u(s_1^\mu)$, contained in a level set of f^μ, for μ near $\bar{\mu}$.

Then $H^\mu: C^\mu \to \tilde{C}^{h(\mu)}$ is fixed in such a way that:

- $C^\mu \cap W^u(s_1^\mu)$ is sent to $\tilde{C}^{h(\mu)} \cap W^u(\tilde{s}_1^{h(\mu)})$, and $C^\mu \cap W^s(s_2^\mu)$ to $\tilde{C}^{h(\mu)} \cap W^s(\tilde{s}_2^{h(\mu)})$.
- H^μ is compatible with the homeomorphisms on the space of leaves of $\mathcal{F}^u(s_*^\mu)$, for hyperbolic singular points s_*^μ with $W^u(s_*^\mu) \cap W^s(s^\mu) \neq \emptyset$.

The extension of H^μ to a full neighbourhood of the orbit of tangency, and subsequently to all of M, causes no harm.

Remark

In [7] Palis and Takens imposed two additional conditions on the family $\{X^{\mu}\}$. These enabled them to reduce the construction to an invariant manifold of dimension two, which can subsequently be extended to a conjugacy on the whole manifold using stable and unstable foliations. However, violation of these conditions does not cause instability of $\{X^{\mu}\}$, so they are not necessary to avoid bifurcations which only may occur in the presence of at least two parameters.

Bifurcations in two parameter families associated with one orbit of tangency.

We finally describe the additional bifurcations in generic two-parameter families corresponding to the occurrence of an orbit of tangency for which exactly one of the conditions (i) - (vii) is violated in the "mildest way possible". They were not contained in an earlier section, since they only occur in dimension ≥ 3.

Observe that it does not make sense to consider violation of condition (iii). It can be replaced by a similar condition for two parameter families. Condition (iv) also holds for generic two-parameter families, since the set of linear gradients on $\mathbb{R}^2$ having two equal eigenvalues has codimension two. Hence generically a pair of equal contracting eigenvalues at s_1^{μ} does not occur together with an orbit of tangency. Finally violation of (vii) gives rise to at least one orbit of tangency or at least one non-hyperbolic singular point, a situation already treated. So we have four different types of bifurcations corresponding to violation of exactly one of the remaining conditions. To describe them successively we consider again a continuous two-parameter family of discs D_r^{μ}, transversal to the orbit of tangency.

1. $\dim W^u(s_1^{\bar{\mu}}) = \dim W^s(s_2^{\bar{\mu}}) = 1$ (violation of (i)).

Secondary bifurcations may be present due to occurrence of an orbit of tangency between $W^s(s_2^{\mu})$ and an unstable manifold (of dimension two) intersecting $W^s(s_1^{\mu})$ or between $W^u(s_1^{\mu})$ and a stable manifold (of dimension two) intersecting $W^u(s_2^{\mu})$. In Figure 10 intersections of D_r^{μ} and unstable manifolds having non-empty intersection with $W^s(s_1^{\mu})$ are labeled $U_1^{\mu},..,U_n^{\mu}$. Similarity $S_1^{\mu},..,S_m^{\mu}$ denote intersections of D_r^{μ} and stable manifolds meeting $W(s_2^{\mu})$
The corresponding bifurcation diagram consist of n + m curves in the parameter plane, having $\bar{\mu}$ in their closure. For parameter values μ on these curves we have

either: $W^s(s_2^{\mu}) \cap D_r^{\mu} \in U_i^{\mu}$, for some i, $1 \leq i \leq n$

or : $W^u(s_1^{\mu}) \cap D_r^{\mu} \in S_i^{\mu}$, for some i, $1 \leq i \leq m$.

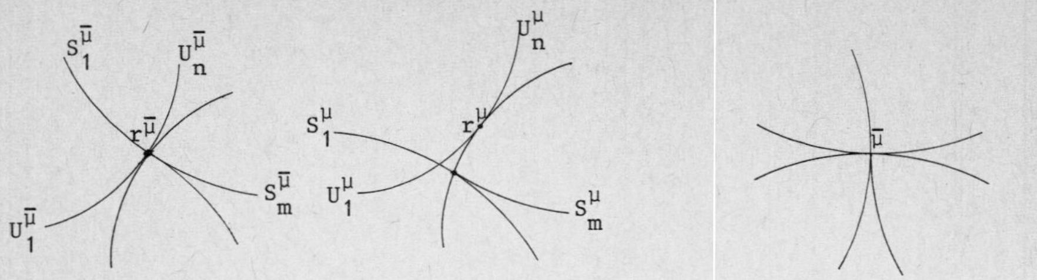

Central bifurcation A secondary bifurcation The bifurcation diagram

Figure 10

2. $\frac{d^2F^{\bar{\mu}}}{dx^2}(0)=0$ (violation of (ii))

For an open and dense class of two-parameter families we have $\frac{d^3F^{\bar{\mu}}}{dx^3}(0) \neq 0$, while the family $\{F^{\mu}\}$ is a 'versal unfolding' of $F^{\bar{\mu}}$, cf. [1]. The latter condition implies existence of local coordinates (μ,x) near $(\bar{\mu},r)$ in which $(\bar{\mu},r)$ corresponds to $(0,0) \in \mathbb{R}^2 \times \mathbb{R}$, such that $F^{\mu}(x) = x^3 + \mu_1 x + \mu_2$. This leads to the bifurcation diagram depicted in Figure 11.

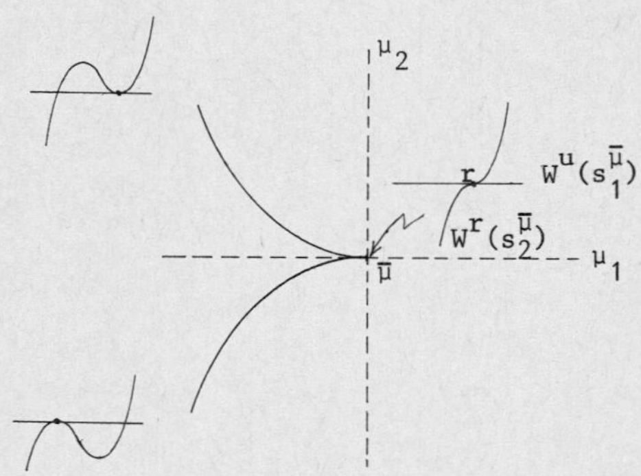

Figure 11

3. Situation of case II, where $\lim_{t\to-\infty} dX_t^{\bar{\mu}}(E_r)$ is the eigenspace of the linear part at $s_1^{\bar{\mu}}$ corresponding to the <u>weakest</u> contracting eigenvalue (voliation of (v)). The situation in D_r^{μ} is depicted in Figure 12. Again D_r^{μ} contains $U_1^{\mu}, \ldots, U_n^{\mu}$ that are intersections of D_r^{μ} and unstable manifolds meeting $W^s(s_1^{\mu})$. Secondary bifurcations occur for parameter values μ for which $W^s(s_2^{\mu}) \cap D_2^{\mu}$ is tangent to U_i^{μ}, for some i, $1 \leq i \leq n$.

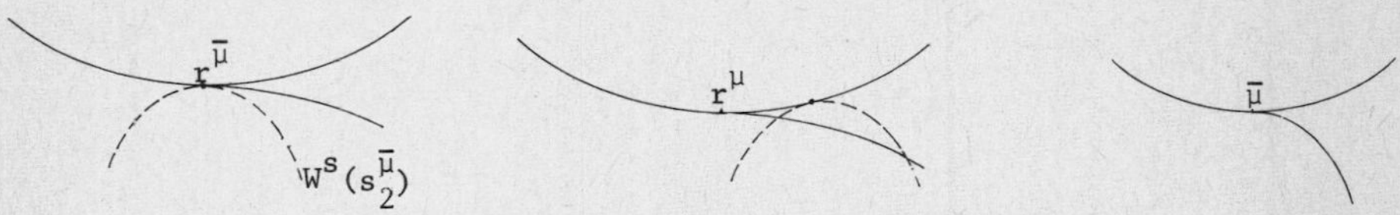

Central bifurcation | A secondary bifurcation | Bifurcation diagram

Figure 12

4. X^{μ} has an orbit of tangency as in case II and exactly one hyperbolic saddle $s_*^{\bar{\mu}}$, different from $s_1^{\bar{\mu}}$ and $s_2^{\bar{\mu}}$, such that $W^u(s_*^{\bar{\mu}})$ and $W^{ss}(s_1^{\bar{\mu}})$ are not transversal. In this case we have secondary bifurcations for parameter values μ for which one of the following situations occur:

I $W^u(s_1^{\mu}) \cap D_r^{\mu} \in W^s(s_2^{\mu}) \cap D_r^{\mu}$

II $W^u(s_*^{\mu}) \cap D_r^{\mu}$ is tangent to $W^s(s_2^{\mu}) \cap D_r^{\mu}$, see Figure 13.

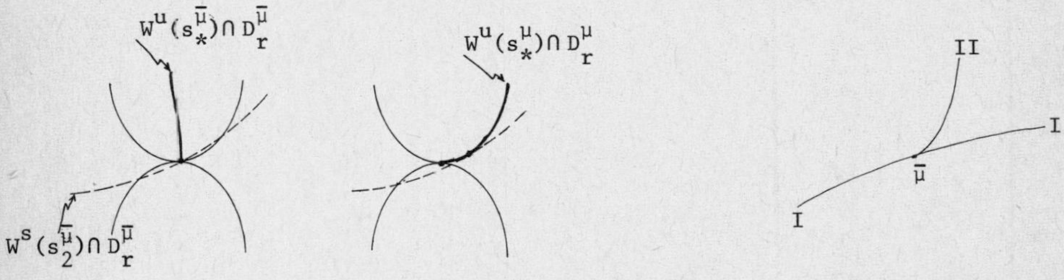

Central bifurcation | Secondary bifurcation (II) | Bifurcation diagram

Figure 13

This completes our description of possible bifurcations of generic two-parameter families of gradients. Applying techniques similar to those described in the preceding section we are able to prove structural stability of these families. The construction of topological equivalences is quite delicate in some cases (especially in case 3, corresponding to figure 12), and it is not obvious that our methods may also be applied if the number of dimensions exceeds three.

References

[1] Bröcker, Th. Differentiable germs and catastrophes, Cambridge University Press, (1975).

[2] Hirsch, M.W., Pugh, C.C., Shub, M. Invariant Manifolds, Springer Lecture Notes in Mathematics 583, (1977).

[3] Newhouse, S., Palis, J., Takens, F. Bifurcations and Stability of Families of Diffeomorphisms, Publications Mathématiques I.H.E.S., 57, (1981).

[4] Palis, J., de Melo, W. Geometric Theory of Dynamical Systems, An Introduction, Springer Verlag, (1982).

[5] Palis, J. A differentiable invariant of topological conjugacies and moduli of stability, Astérisque, 51, 335-346, (1978).

[6] Palis, J., Smale, S. Structural Stability theorems, in: Global Analysis, Proc. A.M.S. Symp. in Pure Math. XIV, 223-232, (1970).

[7] Palis, J., Takens, F. Stability of Parametrized Families of Gradient Vector Fields, Annals of Mathematics, 118, 383-421, (1983).

[8] Takens, F. Moduli of stability for gradients: generic k-parameter families are not always structurally stable, to appear in the Proceedings of a Conference on Dynamical Systems in Heraklion, Crete, ed. S.N. Pneumatikos, (1983).

[9] Vegter, G. Bifurcations of gradient vector fields, Astérisque vol. 98-99, 39-73, (1982).

[10] Vegter, G. Bifurcations of gradient vector fields, Ph.D. Thesis, Groningen, (1983).

Author's address:

Department of Mathematics and
Computing Science
Rijksuniversiteit Groningen
P.O. Box 800
9700 AV Groningen
The Netherlands

Vol. 1034: J. Musielak, Orlicz Spaces and Modular Spaces. V, 222 pages. 1983.

Vol. 1035: The Mathematics and Physics of Disordered Media. Proceedings, 1983. Edited by B.D. Hughes and B.W. Ninham. VII, 432 pages. 1983.

Vol. 1036: Combinatorial Mathematics X. Proceedings, 1982. Edited by L.R.A. Casse. XI, 419 pages. 1983.

Vol. 1037: Non-linear Partial Differential Operators and Quantization Procedures. Proceedings, 1981. Edited by S.I. Andersson and H.-D. Doebner. VII, 334 pages. 1983.

Vol. 1038: F. Borceux, G. Van den Bossche, Algebra in a Localic Topos with Applications to Ring Theory. IX, 240 pages. 1983.

Vol. 1039: Analytic Functions, Błażejewko 1982. Proceedings. Edited by J. Ławrynowicz. X, 494 pages. 1983

Vol. 1040: A. Good, Local Analysis of Selberg's Trace Formula. III, 128 pages. 1983.

Vol. 1041: Lie Group Representations II. Proceedings 1982–1983. Edited by R. Herb, S. Kudla, R. Lipsman and J. Rosenberg. IX, 340 pages. 1984.

Vol. 1042: A. Gut, K. D. Schmidt, Amarts and Set Function Processes. III, 258 pages. 1983.

Vol. 1043: Linear and Complex Analysis Problem Book. Edited by V.P. Havin, S.V. Hruščëv and N.K. Nikol'skii. XVIII, 721 pages. 1984.

Vol. 1044: E. Gekeler, Discretization Methods for Stable Initial Value Problems. VIII, 201 pages. 1984.

Vol. 1045: Differential Geometry. Proceedings, 1982. Edited by A.M. Naveira. VIII, 194 pages. 1984.

Vol. 1046: Algebraic K-Theory, Number Theory, Geometry and Analysis. Proceedings, 1982. Edited by A. Bak. IX, 464 pages. 1984.

Vol. 1047: Fluid Dynamics. Seminar, 1982. Edited by H. Beirão da Veiga. VII, 193 pages. 1984.

Vol. 1048: Kinetic Theories and the Boltzmann Equation. Seminar, 1981. Edited by C. Cercignani. VII, 248 pages. 1984.

Vol. 1049: B. Iochum, Cônes autopolaires et algèbres de Jordan. VI, 247 pages. 1984.

Vol. 1050: A. Prestel, P. Roquette, Formally p-adic Fields. V, 167 pages. 1984.

Vol. 1051: Algebraic Topology, Aarhus 1982. Proceedings. Edited by I. Madsen and B. Oliver. X, 665 pages. 1984.

Vol. 1052: Number Theory. Seminar, 1982. Edited by D.V. Chudnovsky, G.V. Chudnovsky, H. Cohn and M.B. Nathanson. V, 309 pages. 1984.

Vol. 1053: P. Hilton, Nilpotente Gruppen und nilpotente Räume. V, 221 pages. 1984.

Vol. 1054: V. Thomée, Galerkin Finite Element Methods for Parabolic Problems. VII, 237 pages. 1984.

Vol. 1055: Quantum Probability and Applications to the Quantum Theory of Irreversible Processes. Proceedings, 1982. Edited by L. Accardi, A. Frigerio and V. Gorini. VI, 411 pages. 1984.

Vol. 1056: Algebraic Geometry. Bucharest 1982. Proceedings, 1982. Edited by L. Bădescu and D. Popescu. VII, 380 pages. 1984.

Vol. 1057: Bifurcation Theory and Applications. Seminar, 1983. Edited by L. Salvadori. VII, 233 pages. 1984.

Vol. 1058: B. Aulbach, Continuous and Discrete Dynamics near Manifolds of Equilibria. IX, 142 pages. 1984.

Vol. 1059: Séminaire de Probabilités XVIII, 1982/83. Proceedings. Edité par J. Azéma et M. Yor. IV, 518 pages. 1984.

Vol. 1060: Topology. Proceedings, 1982. Edited by L. D. Faddeev and A.A. Mal'cev. VI, 389 pages. 1984.

Vol. 1061: Séminaire de Théorie du Potentiel. Paris, No. 7. Proceedings. Directeurs: M. Brelot, G. Choquet et J. Deny. Rédacteurs: F. Hirsch et G. Mokobodzki. IV, 281 pages. 1984.

Vol. 1062: J. Jost, Harmonic Maps Between Surfaces. X, 133 pages. 1984.

Vol. 1063: Orienting Polymers. Proceedings, 1983. Edited by J.L. Ericksen. VII, 166 pages. 1984.

Vol. 1064: Probability Measures on Groups VII. Proceedings, 1983. Edited by H. Heyer. X, 588 pages. 1984.

Vol. 1065: A. Cuyt, Padé Approximants for Operators: Theory and Applications. IX, 138 pages. 1984.

Vol. 1066: Numerical Analysis. Proceedings, 1983. Edited by D.F. Griffiths. XI, 275 pages. 1984.

Vol. 1067: Yasuo Okuyama, Absolute Summability of Fourier Series and Orthogonal Series. VI, 118 pages. 1984.

Vol. 1068: Number Theory, Noordwijkerhout 1983. Proceedings. Edited by H. Jager. V, 296 pages. 1984.

Vol. 1069: M. Kreck, Bordism of Diffeomorphisms and Related Topics. III, 144 pages. 1984.

Vol. 1070: Interpolation Spaces and Allied Topics in Analysis. Proceedings, 1983. Edited by M. Cwikel and J. Peetre. III, 239 pages. 1984.

Vol. 1071: Padé Approximation and its Applications, Bad Honnef 1983. Prodeedings. Edited by H. Werner and H.J. Bünger. VI, 264 pages. 1984.

Vol. 1072: F. Rothe, Global Solutions of Reaction-Diffusion Systems. V, 216 pages. 1984.

Vol. 1073: Graph Theory, Singapore 1983. Proceedings. Edited by K.M. Koh and H.P. Yap. XIII, 335 pages. 1984.

Vol. 1074: E.W. Stredulinsky, Weighted Inequalities and Degenerate Elliptic Partial Differential Equations. III, 143 pages. 1984.

Vol. 1075: H. Majima, Asymptotic Analysis for Integrable Connections with Irregular Singular Points. IX, 159 pages. 1984.

Vol. 1076: Infinite-Dimensional Systems. Proceedings, 1983. Edited by F. Kappel and W. Schappacher. VII, 278 pages. 1984.

Vol. 1077: Lie Group Representations III. Proceedings, 1982–1983. Edited by R. Herb, R. Johnson, R. Lipsman, J. Rosenberg. XI, 454 pages. 1984.

Vol. 1078: A.J.E.M. Janssen, P. van der Steen, Integration Theory. V, 224 pages. 1984.

Vol. 1079: W. Ruppert. Compact Semitopological Semigroups: An Intrinsic Theory. V, 260 pages. 1984

Vol. 1080: Probability Theory on Vector Spaces III. Proceedings, 1983. Edited by D. Szynal and A. Weron. V, 373 pages. 1984.

Vol. 1081: D. Benson, Modular Representation Theory: New Trends and Methods. XI, 231 pages. 1984.

Vol. 1082: C.-G. Schmidt, Arithmetik Abelscher Varietäten mit komplexer Multiplikation. X, 96 Seiten. 1984.

Vol. 1083: D. Bump, Automorphic Forms on GL (3,IR). XI, 184 pages. 1984.

Vol. 1084: D. Kletzing, Structure and Representations of Q-Groups. VI, 290 pages. 1984.

Vol. 1085: G.K. Immink, Asymptotics of Analytic Difference Equations. V, 134 pages. 1984.

Vol. 1086: Sensitivity of Functionals with Applications to Engineering Sciences. Proceedings, 1983. Edited by V. Komkov. V, 130 pages. 1984

Vol. 1087: W. Narkiewicz, Uniform Distribution of Sequences of Integers in Residue Classes. VIII, 125 pages. 1984.

Vol. 1088: A.V. Kakosyan, L.B. Klebanov, J.A. Melamed, Characterization of Distributions by the Method of Intensively Monotone Operators. X, 175 pages. 1984.

Vol. 1089: Measure Theory, Oberwolfach 1983. Proceedings. Edited by D. Kölzow and D. Maharam-Stone. XIII, 327 pages. 1984.